Emergence, Complexity and Computation

Volume 16

About this Series

The Emergence, Complexity and Computation (ECC) series publishes new developments, advancements and selected topics in the fields of complexity, computation and emergence. The series focuses on all aspects of reality-based computation approaches from an interdisciplinary point of view especially from applied sciences, biology, physics, or chemistry. It presents new ideas and interdisciplinary insight on the mutual intersection of subareas of computation, complexity and emergence and its impact and limits to any computing based on physical limits (thermodynamic and quantum limits, Bremermann's limit, Seth Lloyd limits...) as well as algorithmic limits (Gödel's proof and its impact on calculation, algorithmic complexity, the Chaitin's Omega number and Kolmogorov complexity, non-traditional calculations like Turing machine process and its consequences,...) and limitations arising in artificial intelligence field. The topics are (but not limited to) membrane computing, DNA computing, immune computing, quantum computing, swarm computing, analogic computing, chaos computing and computing on the edge of chaos, computational aspects of dynamics of complex systems (systems with self-organization, multiagent systems, cellular automata, artificial life,...), emergence of complex systems and its computational aspects, and agent based computation. The main aim of this series it to discuss the above mentioned topics from an interdisciplinary point of view and present new ideas coming from mutual intersection of classical as well as modern methods of computation. Within the scope of the series are monographs, lecture notes, selected contributions from specialized conferences and workshops, special contribution from international experts.

More information about this series at http://www.springer.com/series/10624

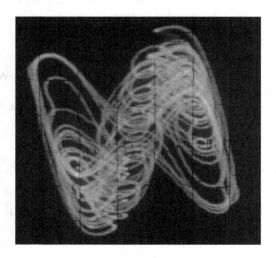

Bharathwaj Muthuswamy · Santo Banerjee

A Route to Chaos Using FPGAs

Volume I: Experimental Observations

 Springer

Bharathwaj Muthuswamy
Software Engineering
Tarana Wireless
Berkeley, CA
USA

Santo Banerjee
Institute for Mathematical Research
  (INSPEM)
University Putra Malaysia
Serdang
Malaysia

ISSN 2194-7287   ISSN 2194-7295   (electronic)
Emergence, Complexity and Computation
ISBN 978-3-319-37537-3   ISBN 978-3-319-18105-9   (eBook)
DOI 10.1007/978-3-319-18105-9

Printed on acid-free paper

Springer International Publishing AG Switzerland is part of Springer Science+Business Media
(www.springer.com)

To the pursuers of knowledge...

Foreword

During the last three decades, the Field-Programmable Gate Array (FPGA) has attracted a lot of interest in the stream of Engineering and Science. The most important reasons for this are the inherent reconfigurability and cost-effectiveness of the FPGA. This device is basically a massively parallel blank slate of hardware (look-up tables, flip-flops, digital signal processing blocks, etc.). The purpose of this book is to configure that hardware to investigate chaotic phenomena.

Even though the FPGA approach is a digital emulation, many complicated nonlinear ODEs can be implemented much more quickly on an FPGA, instead of using traditional analog circuit components such as op-amps and multipliers. For instance, it would take a user around ten minutes to specify and synthesize the Lorenz system on an FPGA. The approach involves drawing a block diagram representation of the Lorenz system nonlinearities in Simulink. Thus it would take the same amount of time to perform a numerical simulation. Nevertheless, proto-typing the Lorenz system on a breadboard using analog multipliers will definitely take more than ten minutes! In addition, compared to traditional microcontrollers, FPGAs are not executing sequential C code. Rather, they are configured specifically for the hardware required.

Moreover, the wide flexibility and the effectiveness of hardware description languages such as VHDL and Verilog enable one to specify long delay chains for FPGA implementation. This is attractive because one can investigate chaotic delay differential equations (DDEs) on FPGAs with ease and even first-order DDEs can be infinite dimensional. Therefore, chaotic DDEs find applications in the field of secure communication.

This book clearly explains the implementation of classic (such as Chen, Chua, Lorenz, Rössler, Sprott) chaotic systems and also chaotic DDEs. For instance, the authors have the very first physical implementation of the Ikeda chaotic DDE. This is possible because the sinusoidal trigonometric nonlinearity required for the DDE can be very easily specified on the FPGA. The authors also leverage the fact that FPGA boards have digital-to-analog converters to view the chaotic waveforms on an oscilloscope.

This volume is very timely since there has been no book so far that describes how to use FPGAs for studying chaotic phenomena. This book should become indispensable for anyone who wants to implement and study chaotic systems, using state-of-the-art FPGA development platforms.

London   Leon O. Chua
February 2015

Preface

If a picture is worth a thousand words, how much is a video worth? To answer this question, please take a look at a video demonstrating the capabilities of a Field-Programmable Gate Array (FPGA) for implementing chaotic systems: https://www.youtube.com/watch?v=wwa7aylrLGo&index=1&list=PLr9kJRBrySkf3P4yiWAxz-CdzaWhTof-PW.

Now that we have you convinced about the robustness of twenty-first century FPGAs, let us talk about the book. In a single sentence, the purpose of this volume is to expose the reader to the exciting world of chaos (science) using FPGAs (engineering). This book differs from other texts on chaos because of a variety of reasons. First, our experimental approach toward chaos theory attempts to provide the reader with a "physical-feel" for the mathematics involved. Second, the medium that we use for physical realization and experiments is the FPGA. These devices are massively parallel digital architectures that can be configured to realize a variety of logic functions. Hence unlike microcontrollers that run sequential compiled code, FPGAs can be configured to execute systems of discrete difference equations in parallel. Moreover, ever since the early twenty-first century, one could realize design specifications in a mathematical simulation software such as Simulink directly onto an FPGA. Also, twenty-first century FPGA development boards have digital-to-analog converters, hence the signals viewed on the oscilloscope are analog waveforms from our digital chaotic realization!

Nevertheless, maximizing the capabilities of an FPGA requires the user to deeply understand the underlying hardware and also the software used to realize the differential equations. This is achieved by another feature in this book: a lab component (along with exercises) in each chapter. In the lab component, readers are asked to investigate chaotic phenomena via MATLAB (or Simulink), design digital logic systems on FPGAs, and also implement a variety of chaotic systems. The specific objective of the lab depends obviously on the particular chapter. Note that one could use FPGAs and development platforms from other manufacturers to understand the concepts in this book. But, for the sake of brevity, we use the Altera Cyclone IV FPGA on a Terasic DE2-115 board which includes an onboard ADA (Analog-to-Digital and Digital-to-Analog) from the audio codec (coder/decoder).

Details on procuring hardware are in Chap. 2. However, understand that FPGA technology is changing rapidly and the hardware (software) used in this book will become quickly outdated. Thus from this book, one has to learn the concepts used in implementing nonlinear ordinary differential equations on FPGAs.

This text is intended for final-year undergraduate or graduate electrical engineering students who are interested in a scientific application of an engineered product. Knowledge of digital logic system (combinational and sequential) realization on FPGAs and an integral calculus course is necessary. A first-year undergraduate course in FPGA-based digital logic and a first-year undergraduate calculus course is necessary and sufficient. However, the only prerequisite for understanding this book is a thirst for knowledge and the willingness to overcome failure. To quote Albert Einstein, "Anyone who has never made a mistake has never tried anything new."

This book is organized as follows: Chapter 1 is an introduction to both chaos theory and FPGAs. Some mathematical foundations for working with chaos are discussed. Chapter 2 gives an overview of the FPGA hardware platform and includes tutorials on utilizing the DE2-115. Chapter 3 shows how to simulate and realize some chaotic ODEs on FPGAs. This chapter combines the ideas in Chaps. 1 and 2 to show the reader how to implement chaotic systems on FPGAs. Chapters 4 and 5 are more mathematical in nature and serve as a precursor to Volume II (Theoretical Methods). Chapter 4 shows how to study bifurcation mechanisms in chaotic systems using FPGAs. Chapter 5 covers time-delayed systems. The FPGA is particularly suited for implementing time-delayed systems because one can implement the necessary delay using n flip-flops (n could be 4096!) by using only five lines in a hardware description language!

There is also a companion website: http://www.harpgroup.org/muthuswamy/ ARouteToChaosUsingFPGAs/ARouteToChaosUsingFPGAs.html for the book that has recorded video solutions (maximum 20 minutes) to all book exercises+labs, FPGA reference designs and corresponding videos (maximum 20 minutes), forums for discussing other hardware platforms, etc. It would be prudent to have access to the internet as you read through this book.

Note that this volume is an "engineering cookbook" that is full of examples for implementing nonlinear dynamical (chaotic) systems on FPGAs. The second volume (theoretical methods) is more rigorous and covers concepts for digital emulation of chaotic dynamics. However, either volumes can be used as stand-alone texts for understanding FPGA-based chaotic system implementation.

This book was typesetted using LATEX. Image processing softwares used were GIMP 2.8, Inkscape 0.48, and Xfig 3.2. An iPhone 5 camera was used for pictures.

NCH Debut Professional v1.64 was used for screen recordings. Oscilloscopes used were the Agilent 54645D and the Tektronix 2205.

Happy chaos via FPGAs!

Santa Clara, California, United States   Bharathwaj Muthuswamy
June 2015   Santo Banerjee

About the book title: it is intended to be a pun on the mathematical concept of a route to chaos.
About the cover art: we show a chaotic attractor from the Ikeda DDE.

Acknowledgments

There are a plethora of folks that we have to thank. From Dr. Muthusamy's perspective, first and foremost, he would like to thank his MS and Ph.D. advisor Dr. Leon O. Chua for all his support and guidance. Ferenc Kovac and Carl Chun from the University of California, Berkeley (Cal) have been both professional and personal mentors throughout the years. Dr. Pravin Varaiya was also instrumental for Dr. Muthuswamy's success at Cal. Dr. Muthuswamy's exposure to cluster computing in 2001 at the Los Alamos National Labs under the guidance of Ron Minnich was invaluable.

This book was technically four years in the making. Dr. Fondjo, Dr. Kennedy, and Ferenc Kovac were extremely helpful in reviewing the very first sample chapters from this book, back in 2010. Faculty colleagues at the Milwaukee School of Engineering, Dr. Jovan Jevtic and Dr. Gerald Thomas, have also provided much needed feedback. Altera (Jeff Nigh, Greg Nash) and Stephen Brown (University of Toronto) donated software and DE2-115 hardware, respectively, along with providing much needed feedback. Many thanks to anonymous peer reviewers from Springer. Their comments were extremely insightful and helpful in fixing errors from the very first version of this book.

Most of the material in this book is based primarily off Dr. Muthuswamy's and Dr. Banerjee's research into FPGA-based nonlinear dynamics. As a result, our research students throughout the years, specifically Ms. Valli from the Vellore Institute of Technology (VIT); Chris Feilbach, Andrew Przybylski from MSOE, and students (Andrew, Curt, Dan, Jonathan, Ryan and Scott) from the very first nonlinear dynamics course offering at MSOE deserve special mention. In addition, Cornell University's ECE 5760 course served as an inspiration for this book. Dr. Muthuswamy was also inspired by Dr. Kennedy et al.'s work on Digital Signal Processor-based investigation of Chua's circuit family (in Chua's Circuit : A Paradigm for Chaos, edited by N. Madan, World Scientific, 1993, pp. 769–792). In many ways, Dr. Muthuswamy's use of FPGAs is a "twenty first century" extension of Dr. Kennedy et al.'s work. Dr. Muthuswamy would also particularly like to thank Ms. Valli, Dr. Ganesan, Dr. C.K. Subramaniam, and others from VIT for providing much needed support in the Summer of 2014 for completing this volume.

A round of applause to Dr. Muthuswamy's spouse, Ms. Deepika Srinivasan, for her help in producing many of the figures. Her "architectural eye" was very helpful in formatting figures so that the appropriate information is properly represented.

Finally, last but not the least, we would like to thank Springer for being very patient with us despite the delay in delivering the final manuscript.

Contents

Acronyms

ADC	Analog-to-Digital Converter
DAC	Digital-to-Analog Converter
DDE	Delay Differential Equation
DSP	Digital Signal Processor/Processing
FPGA	Field-Programmable Gate Array
FSM	Finite-State Machine
HDL	Hardware Description Language
LAB	Logic Array Block
LE	Logic Element
LUT	Look-Up Table
ODE	Ordinary Differential Equation
PLL	Phase-Locked Loop
ROM	Read-Only Memory
RTL	Register Transfer Level
SDC	Synopsys Design Constraints
SDRAM	Synchronous Dynamic Random-Access Memory
SMA	Subminiature Version A
VHDL	Very High Speed Integrated Circuit Hardware Description Language

Mathematical Notations

The mathematical notation used in this book is standard [12]; nevertheless, this section clarifies the notation used throughout the book.

1. Lowercase letters from the Latin alphabet ($a - z$) are used to represent variables, with italic script for scalars and bold invariably reserved for vectors. The letter t is of course always reserved for time. n is usually reserved for the dimension of the state. j is used for $\sqrt{-1}$, in accordance with the usual electrical engineering convention. Mathematical constants such as π, e, h (Planck's constant) have their usual meaning. Other constant scalars are usually drawn from lower case Greek alphabet. SI units are used.
2. Independent variable in functions and differential equations is time (unless otherwise stated) because physical processes change with time.
3. Differentiation is expressed as follows. Time derivatives use Leibniz's ($\frac{dy}{dx}$, for example) or Newton's notation: one, two, or three dots over a variable corresponds to the number of derivatives and a parenthetical superscripted numeral for higher derivatives. Leibniz's notation is used explicitly for non-time derivatives.
4. Real-valued functions, whether scalar- or vector-valued, are usually taken (as conventionally) from lowercase Latin letters f through h, r and s along with x through z.
5. Vector-valued functions and vector fields are bold-faced as well, the difference between the two being indicated by the argument font; hence $\mathbf{f}(x)$ and $\mathbf{f}(\mathbf{x})$ respectively.
6. Constant matrices and vectors are represented with capital and lowercase letters, respectively, from the beginning of the Latin alphabet. Vectors are again bolded.
7. In the context of linear time-invariant systems the usual conventions are respected: A is the state matrix $B(b)$ is the input matrix (vector).
8. Subscripts denote elements of a matrix or vector: $\mathbf{d}_i$ is the *ith* column of D; x_j is the *jth* element of $\mathbf{x}$. Plain numerical superscripts on the other hand may indicate exponentiation, a recursive operation or simply a numbering depending

on context. A superscripted T indicates matrix transpose. I is reserved for the identity matrix. All vectors are assumed to be columns.

9. Σ_i is used for summations, sampling interval is symbolized by T and $\in$ denotes set inclusions.

10. Calligraphic script ($\mathscr{R}$ etc.) is reserved for sets which use capital letters. Elements of sets are then represented with the corresponding lowercase letter. Excepted are the well-known number sets which are rendered in doublestruck bold: $\mathbb{N}, \mathbb{Z}, \mathbb{Q}, \mathbb{R}$ *and* $\mathbb{C}$ for the naturals, integers, rationals, reals, and complex numbers respectively. The natural numbers are taken to include 0. Restrictions to positive or negative subsets are indicated by a superscripted + or −. The symbol $\triangleq$ is used for definitions. $\forall$ and $\exists$ have the usual meaning of "for all" and "there exists" respectively.

Conventions Used in the Book

Each chapter starts with an epigraph, the purpose is to evoke the intellectual curiosity of the reader. Chapters are divided into sections and subsections for clarity. We have placed the most of the MATLAB code and VHDL in appendices to avoid disrupting content flow. For MATLAB code, VHDL and ModelSim scripts, we use line numbers for ease of code discussion.

```
1. MATLAB code or VHDL goes here
```
MATLAB, FPGA Design Suite (Quartus, ModelSim, etc.) menu actions along with respective library block names and in-text elements of VHDL syntax are indicated by **Boldface** notation. Note, however, that further details for utilizing the toolsets are incorporated in online videos on the companion website.

We have used UPPERCASE to describe FPGA pins. The distinction between pins and acronyms will be clear from the context.

Figures and equations are numbered consecutively. The convention for a definition is shown below.

Definition 1 Definitions are typeset as shown.

The book has a variety of solved examples in light gray shade.

Solved Examples

All references are placed at the end of each chapter for convenience. We use a number surrounded by square brackets for in-text references, example [5]. Occasionally, important terminology and concepts are highlighted using **red** font. Although references are hyperlinked, only online URLs are colored **midnight blue** for clarity.

The chapter concludes with a comprehensive set of exercises and a lab.

On a concluding remark, when you find typos in the book please contact the authors with constructive comments: muthuswamy@msoe.edu, santoban@gmail.com.

Chapter 1
Introduction

Lorenz, E.N. *Deterministic*
Nonperiodic Flow [6]

Abstract This chapter will provide a historical overview of chaos and FPGAs. We will begin with a history of how chaos was observed (but unidentified) in a problem related to astronomy and made its way into electronics. On the flip side, the history of FPGAs is a part of the history of Silicon Valley. Next we will look at some very important and fundamental concepts: linearity versus nonlinearity, equilibrium points and Jacobi linearization. As you read through the chapter and work through the exercises, you will realize that nonlinear systems have "rich behaviour" compared to linear systems. Yet you will also notice that relatively simple nonlinear systems can give rise to this rich behaviour.

1.1 An Introduction to Chaos

In this section, we will first go through a very brief history of chaos. The purpose is to emphasize that chaos is essentially a scientific phenomenon that has been studied by engineering for potential applications. The application we will consider is synchronization, although chaos has been used in robot locomotion [1].

1.1.1 A Brief History of Chaos

Chaos was first observed in the late 19th century [2] via a problem related to astronomy! Henri Poincaré, a professor at the University of Paris, observed chaos in the circular restricted three body problem: three masses interacting with each other (think

© Springer International Publishing Switzerland 2015
B. Muthuswamy and S. Banerjee, *A Route to Chaos Using FPGAs*, Emergence, Complexity and Computation 16, DOI 10.1007/978-3-319-18105-9_1

Sun-Earth-Moon). Poincaré determined that it was impossible to accurately predict long term motion of the masses. He discovered the crucial ideas of stable and unstable manifolds which are special curves in the plane. However, even after Poincaré's seminal work, chaos was largely forgotten till the 1920s.

Edwin H. Armstrong invented the regenerative circuit for high frequency oscillations in 1912 [3]. It is possible that he actually observed chaos [3]. In 1927, Dutch physicists Van Der Pol and Van Der Mark proposed one of the first nonlinear oscillators [4, 5]. They observed that at certain drive frequencies, an "irregular noise" is present. These were one of the first discovered instances of deterministic chaos in circuits.

In the late 1950s, a meteorologist at MIT named Edward Lorenz acquired a Royal-McBee LGP-30 computer with 16 kilobyte (KB) of internal memory [2]. Although Lorenz initially started with a system of twelve ordinary differential equations, he eventually simplified the model to a system of three ordinary coupled nonlinear differential equations now known as the Lorenz equations [6], shown in Eqs. (1.1)–(1.3).

$$\dot{x} = -\sigma x + \sigma y \tag{1.1}$$

$$\dot{y} = -xz + \rho x - y \tag{1.2}$$

$$\dot{z} = xy - \beta z \tag{1.3}$$

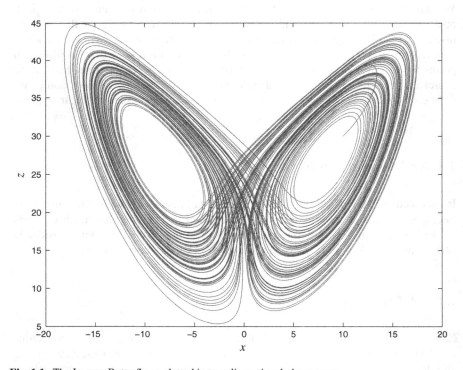

Fig. 1.1 The Lorenz Butterfly as plotted in two dimensional phase space

Notice that since our notation implies that the independent variable in differential equations is time, we have omitted t in the differential equations for clarity. The equations are said to be coupled because the right-hand-side (RHS) of one equation involves a variable from another equation. For example, the RHS of Eq. (1.1) involves $y(t)$, which is the solution to Eq. (1.2). The exact definition of nonlinearity will be clarified later in this chapter.

Figure 1.1 shows the result of simulating Eqs. (1.1)–(1.3) in MATLAB. The code for simulation is shown in listing B.1. You should type the code and reproduce the plot, refer to Appendix A at the end of the book for a short tutorial on MATLAB (and Simulink). The parameter values are $\sigma = 10$, $\rho = 28$, $\beta = \frac{8}{3}$. Initial conditions are $x(0) = 10$, $y(0) = 20$, $z(0) = 30$.

Figure 1.1 is the phase plot obtained from our system. A phase plot is a parametric plot, in this case the x-axis function is $x(t)$ and the y-axis function is $z(t)$. In other words, each point on the plot of the object in Fig. 1.1 corresponds to a solution $(x(t), z(t))$. Figure 1.2 shows a parametric three dimensional plot. We have also plotted the time domain waveform $x(t)$ in Fig. 1.3. The waveform is not visually appealing as Figs. 1.1 and 1.2. For instance, the elegance of the phase plots is absent in the time domain waveform.

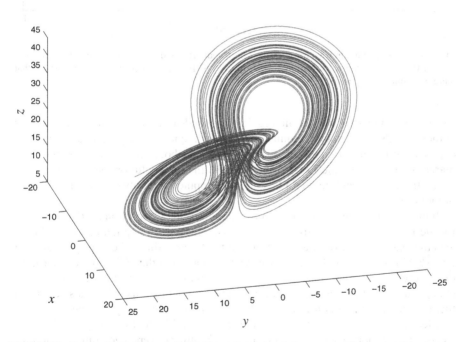

Fig. 1.2 The Lorenz Butterfly as plotted in three dimensional phase space. Comparing this picture to a true three dimensional object such as a sphere, you can see that there is no "inside" and "outside" to the Lorenz Butterfly. This in turn implies that the Lorenz Butterfly has a fractional dimension. In other words, the Lorenz Butterfly is a fractal

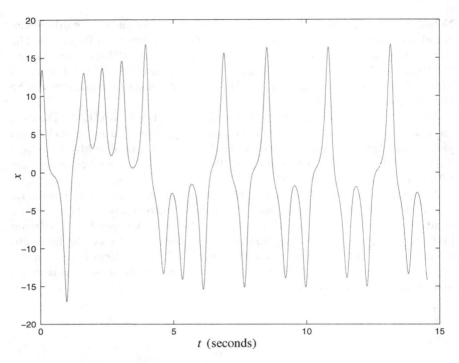

Fig. 1.3 $x(t)$ time domain waveform for the Lorenz butterfly. We have plotted only one waveform for clarity. Although the waveform looks periodic, a fast Fourier transform will clearly show that there is no underlying period

The structure that we see in Fig. 1.2 is called a chaotic strange attractor (or attractor).[1] Using computer simulations, Lorenz identified the presence of sensitive dependence on initial conditions, aperiodicity and bounded trajectories, the hallmarks of chaos. Just like Lorenz, we also obtained the solution using a numerical simulation. The exception might be that our computer probably does not have 16KB of memory.

It turns out that we cannot determine an explicit chaotic solution $(x(t), y(t), z(t))$ for the Lorenz system in Eqs. (1.1)–(1.3). This lack of an explicit chaotic solution lead to a natural question: was the strange attractor an artifact of computer round-off errors? Although other chaotic sytems were proposed in the 1970s and early 1980s, they were also studied through computer simulations. Hence the question of the strange attractor being an artifact of computer simulation was still unanswered and there was a need to demonstrate that chaos is a robust physical phenomenon.

Electronic circuits were a natural choice for realizing the Lorenz system because electronic components such as operational amplifiers were becoming ubiquitous

---

[1]There exist nonchaotic strange attractors and chaotic nonstrange attractors. We will primarily discuss chaotic strange attractors in this book.

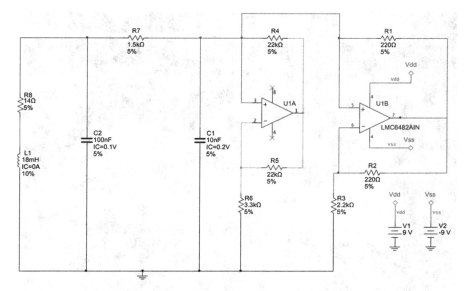

Fig. 1.4 Chua's circuit, the nonlinear resistor is realized using two operational amplifiers

in the early 1970s. However the difficulty in realizing Lorenz and other similar systems via electronic circuits was the fact that the equations describing the chaotic system involved multiplying two functions. The analog multiplier was not a reliable electronic component in the 1970s and early 1980s.

This unreliablity of the analog multiplier spurred the invention of the first electronic chaotic circuit in 1983, almost 20 years after Lorenz's paper. Leon O. Chua, a professor at the University of California, Berkeley designed the first electronic chaotic circuit, Chua's circuit [7]. The circuit and oscilloscope pictures are shown in Figs. 1.4 and 1.5 respectively [8].

By rescaling the circuit variables v_{C1}, v_{C2} and i_L from Fig. 1.4, we obtain the dimensionless[2] Chua Equations shown in Eqs. (1.4)–(1.6). $\alpha, m_1, m_0, \beta \in \mathbb{R}$ are parameters of the system.

$$\dot{x} = \alpha[y - x - m_1 x - \frac{1}{2}(m_0 - m_1)(|x + 1| - |x - 1|)] \tag{1.4}$$

$$\dot{y} = x - y + z \tag{1.5}$$

$$\dot{z} = -\beta y \tag{1.6}$$

Notice that the circuit in Fig. 1.4 is not an analog computer. That is, we do not have analog integrators for solving the system of equations in (1.4)–(1.6). Chua was able to construct the circuit in Fig. 1.4 without analog integrators because he systematically derived the circuit for producing chaos from basic concepts in nonlinear

---

[2]Dimensionless formulation will be covered in Sect. 4.2.4.

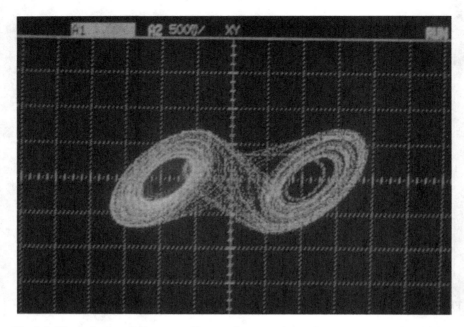

Fig. 1.5 Phase plot recorded on an oscilloscope from experimental measurements of Fig. 1.4. The inductor was realized using an op-amp operating as a gyrator [9]

circuit theory. This approach was also instrumental in mathematically proving the existence of chaos [8] in Chua's circuit. Thus Chua's circuit was the first system in which the existence of chaos was confirmed numerically via simulation, experimentally via electronics and rigorously via Shilnikov's theorem by 1984. Between the announcement of the circuit in late 1983 and the rigorous proof of chaos by 1984, the time span was approximately one year. In comparison, Lorenz's system was rigorously proved to be chaotic only in 1999 by Warwick Tucker, a span of 36 years since Lorenz's paper in 1963! Chua's approach illustrates the paradigm of using electronics to study chaos—we have at our disposal a physical interpretation of chaos. This physical interpretation of chaos is the motivation behind using FPGAs to study the phenomenon.

Since Chua's work, many other chaotic circuits have been proposed. A family of such circuits involve jerky dynamics and were proposed by Julien Clinton Sprott from the University of Wisconsin, Madison [10]. One possible chaotic circuit based on jerky dynamics is shown in Fig. 1.6. An oscilloscope phase plot is shown in Fig. 1.7. Notice that unlike Chua's circuit, Fig. 1.6 is a circuit based on analog integrators. These circuits are easy to build and analyze analytically. We will realize the system equations on FPGAs later in the book.

$$\dddot{x} = J(x, \dot{x}, \ddot{x}) \tag{1.7}$$

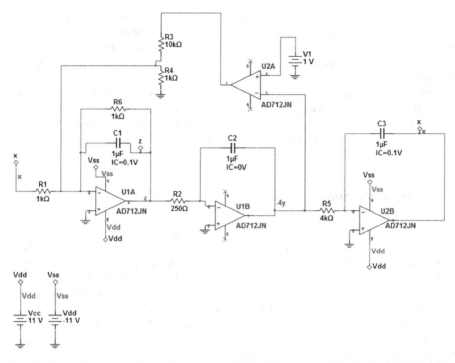

Fig. 1.6 A chaotic circuit realizing jerky dynamics. This circuit was implemented by former MSOE students Chris Feilbach and Clara Backes

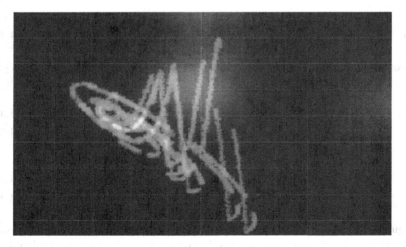

Fig. 1.7 Phase plot from an oscilloscope screenshot for the circuit in Fig. 1.6. $\ddot{x}$ is on the y-axis, x is plotted on the x-axis

Equation (1.7) shows the general system equation for Sprott's jerky dynamical systems. They are so named because if $x(t)$ is considered to be position then Eq. (1.7) implies that $\ddot{x}$ is the acceleration. But Eq. (1.7) involves the derivative of acceleration or the jerk. Jerky dynamics is very useful in rocket science, although we unfortunately will not be building rockets in this book.

After Sprott, a variety of other chaotic circuits (hysteresis based chaos generators, chaos from synchronized oscillators etc.) have been developed. We will leave the history of chaos with Sprott and turn our attention to one application of chaos that will be discussed later in the book—synchronization.

1.1.2 An Application of Chaos

One very interesting application of chaos is synchronization for secure communication: a transmitter and receiver chaotic system can synchronize with each other. But if we use the chaotic signal as a much larger masking signal, then we can transmit a message using the chaotic mask. The concept of synchronization in chaotic systems was originally proposed by Pecora and Carroll [11]. An application to secure communication was suggested by Cuomo and Oppenheim [12].

The key to this concept is that if a chaotic system (say the Lorenz system) can be decomposed into subsystems, a drive subsystem and a stable response subsystem, then the original message can be recovered at the receiver using only the transmitted signal. Consider again the Lorenz system of equations.

$$\dot{x} = -\sigma x + \sigma y \tag{1.8}$$

$$\dot{y} = -xz + \rho x - y \tag{1.9}$$

$$\dot{z} = xy - \beta z \tag{1.10}$$

Pecora and Carroll [11] showed that Eqs. (1.8)–(1.10) can be decomposed into two stable response subsystems.

$$\dot{x}_1 = -\sigma x_1 + \sigma y \tag{1.11}$$

$$\dot{z}_1 = x_1 y - \beta z_1 \tag{1.12}$$

$$\dot{y}_2 = -xz_2 + \rho x - y_2 \tag{1.13}$$

$$\dot{z}_2 = xy_2 - \beta z_2 \tag{1.14}$$

Equations (1.8)–(1.10) can be interpreted as the drive system since its dynamics are independent of the response subsystems. Nevertheless, the two response subsystems (Eqs. (1.11)–(1.14)) can be used together to regenerate the full-dimensional dynamics which are evolving at the drive system [12]. Specifically, if the input signal to the (y_2, z_2) subsystem is $x(t)$, then the output $y_2(t)$ can be used to drive the (x_1, z_1)

subsystem and subsequently generate a "new" $x(t)$ in addition to having obtained, through synchronization, $y(t)$ and $z(t)$.

We will study such synchronization mechanisms in delay differential equations in Chap. 5.

1.2 An Introduction to Field Programmable Gate Arrays

We will now take a digression from science and give a brief overview of a very flexible integrated circuit—the FPGA.

1.2.1 History of FPGAs

The FPGA industry originated from the programmable read-only memory and programmable logic devices industry of the 1970s. Xilinx co-founders Ross Freeman and Bernard Vonderschmitt invented the first commercially viable field programmable gate array in 1985 [13]—the XC2064.[3]

Freeman and Vonderschmitt were both working as chip engineers at Zilog Corp. prior to joining Xilinx. While working at Zilog, Freeman wanted to design a computer chip that effectively acted as a blank tape, allowing the user to program the chip "in hardware" rather than having to purchase a preprogrammed chip (or ASIC—Application Specific Integrated Circuit) from the manufacturer. Freeman approached his superiors at Zilog and suggested that such a programmable chip would be a viable new avenue for the company. Nevertheless, he was unable to convince executives at Exxon (Zilog's parent company) to chase a totally unexplored market. As a result, Freeman left his post at Zilog and along with Vonderschmitt founded Xilinx.

Xilinx's FPGA was based on the company's patented Logic Cell Array technology. The company's system basically consisted of an off-the-shelf programmable chip and a software package that could be used to program and tailor the chip for specific needs. The technology was based on the arrangement of gates (the lowest level building block in a logic circuit) in complex formations called arrays; as the number of gates increased, the more complex were the functions that the semiconductor could perform. Figure 1.8 shows a very simple FPGA Logic Element or LE.

In this book, we will utilize FPGAs from Altera (and a development board from Terasic Inc.). Altera's history is as interesting as Xilinx. We will not discuss their history more except to note that the name "Altera" is from "alterable" [14].

---

[3] Although Xilinx's competitor, Altera, was founded in 1983.

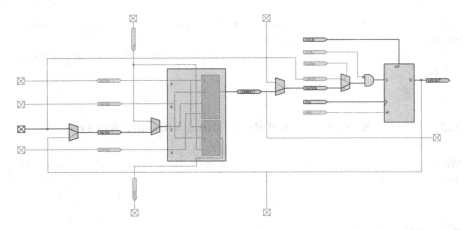

Fig. 1.8 The basic processing unit on an Altera FPGA, the Logic Element (LE). Screenshot has been obtained using Altera's Chip Planner tool in Quartus 12.0

1.2.2 Why FPGAs?

Unlike processors, FPGAs use dedicated hardware for processing logic and hence are not constrained by the complexities of additional overhead, such as an operating system. In the early days of FPGAs, they were usually constrained by high power requirements. Nevertheless with the latest FPGA families (such as Stratix from Altera) emphasizing low dynamic power performance, FPGAs are being increasingly used for digital signal processing (DSP) applications [15].

Another interesting benefit of FPGA technology is that since it is truly a "hard" implementation of our design specification, FPGAs provide more reliability unlike software tools running in a programming environment. This is because processor-based systems often involve several layers of abstraction to help schedule tasks and share resources among multiple processes. All these complexities are unnecessary in an FPGA based system.

From the standpoint of differential equations, one can recast a fixed step algorithm (such as Euler's method) in a simple block diagram form suitable for realization on an FPGA, refer to Fig. 1.9.

However, before we can realize differential equations on FPGAs we need to learn some basic mathematical concepts. That is the subject of the next section.

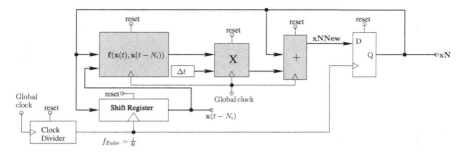

Fig. 1.9 The block diagram that we will eventually implement in this book for solving the nonlinear delay differential equation (DDE): $\dot{x} = f(t, x(t), x(t - \tau_i))$ [16]. The blocks in *grey* will be implemented in MATLAB using the DSP Builder Advanced Blockset from Altera

1.3 Some Basic Mathematical Concepts

We will start with the very basic concept of what is a linear system (and what is not). This will eventually help us understand what kind of systems give rise to chaotic behaviour.

1.3.1 Linear Versus Nonlinear Equations

Consider Eqs. (1.15) and (1.16), a system of linear equations:

$$x - 3y = 3 \tag{1.15}$$

$$2x - 6y = 6 \tag{1.16}$$

What are the solutions to Eqs. (1.15) and (1.16)? To answer this question notice that Eq. (1.16) can be simplified to $x - 3y = 3$. Thus we have only one equation in two unknowns, shown below.

$$y = \frac{x}{3} - 1 \tag{1.17}$$

Hence if we let $x \in \mathbb{R}$, there are infinitely many real number solutions in the form (x, y). In other words, we have two superimposed straight lines in our solution space. The MATLAB code in listing B.2 plots one of the equations, the result is Fig. 1.10.

The beauty of linear equations is that we can have only two other kinds of solutions to a system of linear equations: a unique solution or no solution. The detailed study of linear systems and their application is the subject of linear algebra [17]. We will introduce concepts from linear algebra, when required in this book.

Unlike linear equations, one cannot state beforehand (or *a priori*) how many solutions a nonlinear equation can have in general. Take the case of a simple quadratic

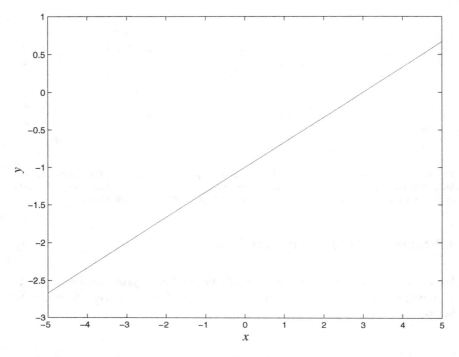

Fig. 1.10 The function $y = \frac{x}{3} - 1$ as plotted in MATLAB

equation: $y = x^2 - 2$. If $x \in \mathbb{R}$ then the equation has two solutions: $x = \pm\sqrt{2}$. The situation changes drastically if we have non-polynomial functions involved.

However we can understand the difference between linearity and nonlinearity using the concept of a system.[4] Loosely speaking a system "acts" on a function. For example consider the volume knob on your car stereo. This volume knob adjusts the gain of your audio amplifier. In other words the audio amplifier multiplies your input signal by a factor proportional to how much you turn the volume knob. Mathematically, the audio amplifier transforms an input function into another output function. Hence we have our first definition [18].

Definition 1.1 A system is a function whose domain and range are sets of signals called signal spaces.

For example, let us say the audio amplifier has a gain $K \in \mathbb{R}$ that can be controlled by the volume knob. Then we can say that the audio amplifier is a system S such that:

$$y = S(x) \tag{1.18}$$

---

[4]Note that we are not talking about a system of equations. This difference should become clear after we provide a mathematical definition of a system.

x and y are functions in the function space of continuous-time systems: $x, y : t \rightarrow \mathbb{R}$. Thus the output of the system at any time t is given by:

$$y(t) = S(x)(t) \tag{1.19}$$
$$= Kx(t) \tag{1.20}$$

Thus if the gain K is positive, then we hear loud (sometimes obnoxiously loud) music.

We need to emphasize that we should not write the mathematical description of the system as $S(x(t))$. This is because a system's domain is a function. $x(t)$ is a number, not a function.

Now that we know the definition of a system, a linear system is defined below [18].

Definition 1.2 A system S is said to be linear iff: $\forall u, v \in \mathcal{D}, \forall \alpha, \beta \in \mathbb{R}, S(\alpha u + \beta v) = \alpha S(u) + \beta S(v)$.

In Definition 1.2, set $\mathcal{D}$ is the domain of S. In other words, Definition 1.2 is the superposition theorem: "response of the sum is the sum of the responses". Let us understand the definition via examples.

Example 1.1 Consider the audio amplifier system: $y(t) = S(x)(t) = Kx(t)$. Is this system linear or nonlinear?

Solution: Applying Definition 1.2 to the audio amplifier system, we have:

$$S(\alpha u + \beta v) = K(\alpha u + \beta v) \tag{1.21}$$
$$= K\alpha u + K\beta v \tag{1.22}$$
$$= \alpha K u + \beta K v \tag{1.23}$$
$$= \alpha S(u) + \beta S(v) \tag{1.24}$$

Thus our audio amplifier system is linear.

Note that in reality you cannot expect to get infinite gain out of your audio amplifier. That is, physically the audio amplifier is going to eventually saturate as you turn the volume knob. However the linearity model is an excellent approximation when the audio amplifier does not saturate and we listen to music in the linearity range of the audio amplifier.

Example 1.2 Consider a system that squares the input function: $y(t) = S(x)(t) = x^2(t)$. Is this system linear or nonlinear?

Solution: Applying Definition 1.2 to the square system, we have:

$$S(\alpha u + \beta v)(t) = (\alpha u + \beta v)^2(t) \tag{1.25}$$

$$= \alpha^2 u^2(t) + \beta^2 v^2(t) + 2\alpha\beta u(t)v(t) \tag{1.26}$$

However,

$$S(\alpha u + \beta v)(t) \neq \alpha u^2(t) + \beta v^2(t) \tag{1.27}$$

$$= \alpha S(u)(t) + \beta S(v)(t) \tag{1.28}$$

Thus our square system is nonlinear.

1.3.2 Linear Versus Nonlinear Dynamics

The systems (scaling and squaring) we studied in Sect. 1.3.1 are examples of non dynamical systems. On the other hand, a dynamical system is governed by differential equations, and such systems are the topic of study in this book. We will be concerned with nth order ordinary nonlinear differential equations (ODE), autonomous or non-autonomous, with or without delay.

Definition 1.3 The order of a differential equation is the order of the highest derivative in the equation.

Example 1.3 What is the order of the differential equation: $(\ddot{x})^3 + \sin(x) = 0$?

Solution: Since the highest derivative appearing in the equation is the 2nd derivative, the order of the differential equation is two.

Definition 1.4 A differential equation involving ordinary (non-partial) derivatives is an ODE.

Definition 1.5 An autonomous ODE does not involve an explicit function of time.

Example 1.4 Is Eq. (1.29) below an autonomous ODE?

$$\ddot{x} + \dot{x} + x = \cos(t) \tag{1.29}$$

Solution: Equation (1.29) involves only ordinary derivatives and hence is an ODE. But it does not involve an explicit function of time so the equation is non-autonomous. In this book, we will use the following change of variables to convert non-autonomous ODEs to autonomous ODEs.

$$x_1 = x \tag{1.30}$$
$$x_2 = \dot{x} \tag{1.31}$$
$$x_3 = t \tag{1.32}$$

Using Eqs. (1.29)–(1.32), we get:

$$\dot{x}_1 = x_2 \tag{1.33}$$
$$\dot{x}_2 = -x_1 - x_2 + \cos(x_3) \tag{1.34}$$
$$\dot{x}_3 = 1 \tag{1.35}$$

The advantage of this change of variables is that it allows us to visualize a phase plot with trajectories "frozen" in it.

Linearity or nonlinearity of differential equations can be determined using superposition from Definition 1.2, as the following example illustrates.

Example 1.5 Consider the jerky dynamical system below.

$$\dot{x} = y \tag{1.36}$$
$$\dot{y} = z \tag{1.37}$$
$$\dot{z} = -x - y - \text{sign}(1 + 4y) \tag{1.38}$$

Prove the system above is nonlinear. sign is the signum function defined below.

$$\text{sign}(x) = \begin{cases} -1 & \text{when } x < 0, \\ 0 & \text{when } x = 0, \\ 1 & \text{when } x > 0 \end{cases} \tag{1.39}$$

Proof First, we will write the system above as one third order differential equation in x. To do so, notice that Eqs. (1.37) and (1.36) imply $\dot{z} = \ddot{y} = \dddot{x}$. Thus substituting $\dddot{x}$ for $\dot{z}$ and $\dot{x}$ for y in Eq. (1.38), we get:

$$\dddot{x} + \dot{x} + \text{sign}(1 + 4\dot{x}) + x = 0 \tag{1.40}$$

Suppose there exist three solutions to Eq. (1.40): x_1, x_2, x_3. That is:

$$\dddot{x}_1 + \dot{x}_1 + \text{sign}(1 + 4\dot{x}_1) + x_1 = 0 \qquad (1.41)$$
$$\dddot{x}_2 + \dot{x}_2 + \text{sign}(1 + 4\dot{x}_2) + x_2 = 0 \qquad (1.42)$$
$$\dddot{x}_3 + \dot{x}_3 + \text{sign}(1 + 4\dot{x}_3) + x_3 = 0 \qquad (1.43)$$

We will apply Definition 1.2 and check if a superposition of the solutions: $\alpha x_1 + \beta x_2 + \delta x_3, \alpha, \beta, \delta \in \mathbb{R}$ is also a solution. That is:

$$\frac{d^3}{dt^3}(\alpha x_1 + \beta x_2 + \delta x_3) + \frac{d}{dt}(\alpha x_1 + \beta x_2 + \delta x_3)$$
$$+ \text{sign}\left(1 + 4\frac{d}{dt}(\alpha x_1 + \beta x_2 + \delta x_3)\right) + (\alpha x_1 + \beta x_2 + \delta x_3) \overset{?}{=} 0$$
$$(1.44)$$

Simplifying the left-hand-side (LHS) of Eq. (1.44), we get:

$$\text{LHS} = \alpha(\dddot{x}_1 + \dot{x}_1 + x_1) + \beta(\dddot{x}_2 + \dot{x}_2 + x_2) + \delta(\dddot{x}_3 + \dot{x}_3 + x_3)+$$
$$\text{sign}\,(1 + 4(\alpha \dot{x}_1) + 4(\beta \dot{x}_2) + 4(\delta \dot{x}_3))$$
$$(1.45)$$

Note that if:

$$\text{sign}\,(1 + 4(\alpha \dot{x}_1) + 4(\beta \dot{x}_2) + 4(\delta \dot{x}_3))$$
$$= \text{sign}(1 + 4\alpha \dot{x}_1) + \text{sign}(1 + 4\beta \dot{x}_2) + \text{sign}(1 + 4\delta \dot{x}_3) \qquad (1.46)$$

then Eq. (1.44) is zero by virtue of Eqs. (1.41)–(1.43). However, the signum function is nonlinear. Equation (1.39) shows that signum returns the sign of the input number x: -1 if x is negative, 0 if x is zero and 1 if x is positive. Sketch the signum function or use a few counter-examples to convince yourself that signum is nonlinear. Thus, Eq. (1.44) may not be zero. Hence Definition 1.2 is not satisfied and this implies that our system is nonlinear.   □

The point to be noted from this example is that a rigorous proof of nonlinearity may involve a bit of work. But a quick glance at Eqs. (1.36)–(1.38) reveals the signum function is the reason our system in nonlinear. In other words, if the RHS of our system of first-order ODEs has a nonlinear function, our system is nonlinear.

There is an interesting subset of linear systems that have the same behaviour independent of time shifts: linear time-invariant systems.

Definition 1.6 A system S is said to be linear time-invariant (or LTI) iff $(S \circ D)(x) = (D \circ S)(x)$.

Definition 1.7 defines a time shift using a delay system.

Definition 1.7 $D_\tau(x)(t) = x(t - \tau)$.

Now that we have studied the differences between linear and nonlinear systems, a logical next step would be to try and find an explicit solution to the system under question. We already know that this may not be possible, case in point being an explicit chaotic solution to the Lorenz system does not exist. However, for some differential equations, it may be possible to find an explicit solution.

Example 1.6 Consider the DDE in Eq. (1.47)

$$\dot{x} = D_1(x)(t) \ History(t) = 1, t \leq 0. \tag{1.47}$$

Find $x(t)$ for $t \geq 0$.

Solution: In order to solve the DDE, we will solve the differential equation over mutually exclusive intervals as shown below.

For $0 \leq t < 1$, the DDE can be written as:

$$\dot{x} = 1 \tag{1.48}$$

Equation (1.48) is valid since $D_1(x)(t) = History(t)$ if $0 \leq t < 1$. Notice also that Eq. (1.48) justifies our choice of the label "History". We have an infinite set of initial conditions in the form of a "History" function. Solving Eq. (1.48),

$$x(t) = t + c_0, \ 0 \leq t < 1 \tag{1.49}$$

Now, c_0 in Eq. (1.49) can be determined because we have defined the value of the history function at $t = 0$: $x(0) = History(0) = 1$. Hence the solution to our DDE in the interval $0 \leq t < 1$ is:

$$x(t) = t + 1, \ 0 \leq t < 1 \tag{1.50}$$

Proceeding in this manner, we see that the solution to our differential equation are polynomials of increasing order:

$$x(t) = \begin{cases} t + 1 & 0 \leq t < 1, \\ \frac{t^2}{2} + t + c_1 & 1 \leq t < 2, \\ \cdots & \cdots \end{cases} \tag{1.51}$$

In order to find c_1 in Eq. (1.51), we will impose continuity constraints:

$$t + 1|_{t=1} = \frac{t^2}{2} + t + c_1|_{t=1} \tag{1.52}$$

Thus $c_1 = \frac{1}{2}$.

Now, contrast the solution to our DDE with the solution that corresponds to a differential equation with no delay: $\dot{x} = x$. The solution to the differential equation with no delay is the exponential function. Quite a contrast to the solution in this example!

The example above shows that we need an infinite set of initial conditions to properly solve a DDE. Thus even first order DDEs are infinite-dimensional and can exhibit chaos (and hyperchaos etc.). In Chap. 5, we will simplement chaotic DDEs.

However most physical systems cannot be solved explicitly. But, interestingly, we can predict the behavior of most physical systems without solving for an explicit closed form solution. An introduction to this approach is the topic of Sect. 1.3.3.

1.3.3 Fixed (Equilibrium) Points

Consider the differential equation in Eq. (1.53), written explicitly using time t:

$$\frac{dx}{dt} = \cos(x(t)) \tag{1.53}$$

Physically, points of interest are $x^\dagger$ for which the derivative $\frac{dx}{dt}\big|_{x^\dagger} = 0$. In other words, the system does not "move" from $x^\dagger$. Such points are called fixed points or equilibrium points.

We can extend the above description to nth order differential equations by rewriting an nth order differential equation as n first-order differential equations. In other words, our definition of equilibrium will involve a system of n first-order differential equations as shown below.

Definition 1.8 A system of differential equations $\dot{\mathbf{x}} = \mathbf{f}(\mathbf{x})$ has equilibrium point(s) $\mathbf{x}^\dagger$ such $\dot{\mathbf{x}}\big|_{\mathbf{x}^\dagger} = 0$. Here, $\mathbf{x} \in \mathbb{R}^n$.

Hence given a system of differential equations, we need to solve a set of nonlinear equations for finding the equilibrium points. Let us look at a couple of examples, starting with Eq. (1.53).

Example 1.7 Determine the equilibrium points for the system in Eq. (1.54).

$$\dot{x} = \cos(x) \tag{1.54}$$

Solution: Applying Definition 1.8, we get:

$$\dot{x}\Big|_{x^\dagger} = 0 \tag{1.55}$$

Thus:

$$\cos(x^\dagger) = 0 \tag{1.56}$$

Hence the equilibrium points are the zeroes of the cosine function:

$$x^\dagger = (2k + 1)\frac{\pi}{2}, \ k \in \mathbb{Z} \tag{1.57}$$

Example 1.8 Determine the equilibrium points for the Lorenz system with the specific value of parameters shown below.

$$\dot{x} = -10x + 10y \tag{1.58}$$
$$\dot{y} = -xz + 28x - y \tag{1.59}$$
$$\dot{z} = xy - \frac{8}{3}z \tag{1.60}$$

Solution: Applying Definition 1.8, we get:

$$-10x^\dagger + 10y^\dagger = 0 \tag{1.61}$$
$$-x^\dagger z^\dagger + 28x^\dagger - y^\dagger = 0 \tag{1.62}$$
$$x^\dagger y^\dagger - \frac{8}{3}z^\dagger = 0 \tag{1.63}$$

Equation (1.61) implies

$$x^\dagger = y^\dagger \tag{1.64}$$

Replacing $x^\dagger$ with $y^\dagger$ in Eqs. (1.62) and (1.63) and solving, we get $y^\dagger$ and $z^\dagger$. Hence our equilibrium points are:

$$(x^\dagger, y^\dagger, z^\dagger) = (0, 0, 0), (6\sqrt{2}, 6\sqrt{2}, 27), (-6\sqrt{2}, -6\sqrt{2}, 27) \tag{1.65}$$

1.3.4 System Behaviour Near Fixed Points

Revisiting the Lorenz system from Sect. 1.3.3, if $(x(0), y(0), z(0)) = (x^\dagger, y^\dagger, z^\dagger)$ then $((x(t \to \infty), y(t \to \infty), z(t \to \infty)) = (x^\dagger, y^\dagger, z^\dagger)$. That is if our initial conditions are exactly equilibrium points, we will continue to stay at the equilibrium point, courtesy of Definition 1.8.

However, physically speaking, there is always some inherent noise present in our initial conditions. So a practical question is: what happens when we start our system "near" (or close to) a fixed point? The answer to the question is given by the stability theory of dynamic systems, a topic that is beyond the scope of this volume. The short answer is: if the equilibrium point is stable, our system trajectories will move back towards the equilibrium point. If the equilibrium point is unstable, the system trajectories will move away from the equilibrium point. The unstable case is the interesting one since it may lead to the chaotic trajectories in phase space.

But we can determine what happens to the system behavior near the equilibrium point via numerical simulation using MATLAB. Let us start with the Lorenz system.

Example 1.9 Simulate the Lorenz system starting with the initial conditions $(x(0), y(0), z(0)) = (8.5, 8.7, 3.0)$. Plot a two dimensional phase plot $(x(t), z(t))$.

Solution: We have already simulated the Lorenz system in the introduction section. For this question, we are just going to change the initial condition. We leave it to you as an exercise to generate Fig. 1.11.

If you recall our Lorenz example from the introduction section, the initial conditions were $(x(0), y(0), z(0)) = (10, 20, 30)$. In this example, we start quite a distance away $(10, 20, 30)$. Yet our system still manages to get into a chaotic state. In other words, the Lorenz attractor is robust. Refer to Problem 1.11 for quantifying "robustness".

For two dimensional systems, you can of course write a MATLAB program for simulating the system. However, a very nice tool is pplane7,[5] available from Rice University [19]. Download pplane7 from [19].

Next, start MATLAB and navigate to the directory where you downloaded type pplane7. Type pplane7 at the prompt and press enter. The window in Fig. 1.12 should appear.

Choose the vibrating spring as the system to simulate: **Gallery** → **vibrating spring**. Leave the default parameter value for d as 0 and left-click "Proceed". The window in Fig. 1.13 should appear.

---

[5]An alternative to pplane is the MATLAB command quiver. We will explore the use of quiver in the lab component of this chapter. Nevertheless, pplane is excellent MATLAB code and is open-source. One is encouraged to explore coding styles used in pplane7.

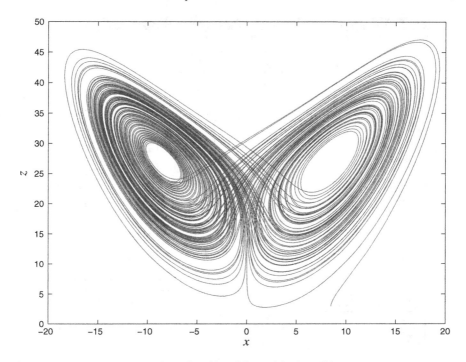

Fig. 1.11 We obtain the Lorenz butterfly with a different initial condition

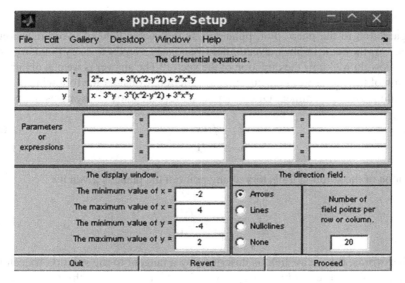

Fig. 1.12 pplane7 startup

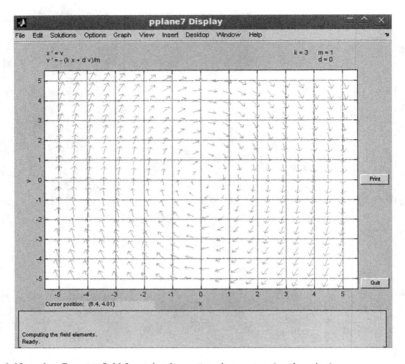

Fig. 1.13 pplane7 vector field for a simple mass-spring system (no damping)

In addition to the phase plot, pplane plots the vector field of two dimensional systems. As the name implies, a vector field associates a vector to every point in phase space. How do we know the direction of the vector? The answer is we can obtain the direction of the vector from the differential equation. Consider the following system of two dimensional nonlinear ODEs. These are representative of systems in pplane.

$$\dot{x} = f_1(x, y) \tag{1.66}$$

$$\dot{y} = f_2(x, y) \tag{1.67}$$

The slope of the solution trajectory at any point in the plane is given by:

$$\frac{dy}{dx} = \frac{\frac{dy}{dt}}{\frac{dx}{dt}} = \frac{f_2(x, y)}{f_1(x, y)} \tag{1.68}$$

In other words, the vector field is described by a tangent to the solution trajectory. Figure 1.14 show a few solution trajectories in pplane. To obtain these trajectories, left-click anywhere in the vector field. Refering to Fig. 1.14, we see that $y(t)$ is $v(t)$ or velocity. Thus to obtain an analytic expression of the vibrating spring vector field, we use Eq. (1.68).

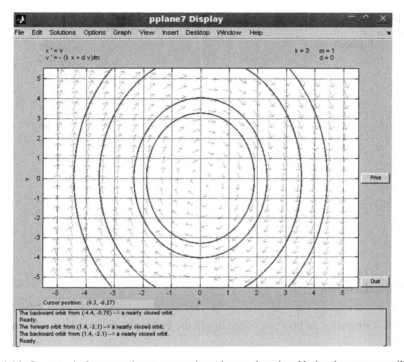

Fig. 1.14 Some typical mass-spring system trajectories, no damping. Notice the system oscillates forever. This system is an ideal case in the sense that most practical (except superconducting) physical systems will have some damping

$$\frac{dy}{dx} = \frac{dv}{dx} = \frac{\frac{-kx}{m}}{v} = \frac{-3x}{v} \tag{1.69}$$

A mechanical schematic of our system is shown in Fig. 1.15. We have a system that is governed by Newton's law and Hooke's law.

$$m\ddot{x} = -kx \tag{1.70}$$

In our case, $m = 1$ kg and $k = 3 \frac{N}{m}$. Using Fig. 1.15 as the physical model for Fig. 1.14, we can further understand the vector field. If we give an initial position and velocity to the mass, then the system oscillates forever since there is no damping. Equation (1.70) is also called the simple harmonic oscillator.

Fig. 1.15 A mass spring
system with the positive
direction of displacement
labelled

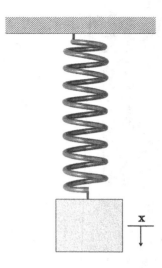

1.4 Conclusions

Below is a summary of how we could apply the concepts and techniques from this
chapter.

1. Given a system of differential equations, determine if they are linear or nonlinear.
2. Compute the equilibrium points[6] of the system using Definition 1.8.
3. Numerically simulate the system using MATLAB and display phase plots. Start
   with initial conditions near the equilibrium points.
4. The phase plot will qualitatively describe system behavior. Use the phase plot as
   a starting point for rigorous analysis.

The beauty of the equilibrium point method lies in the fact that we do not find
an explicit analytic solution. Rather, we determine any equilibrium points and determine system behavior by starting close to those equilibrium points. An example
application of the equilibrium point method was the Lorenz system. We first found
the equilibrium point(s). Then, using MATLAB, we found that starting close to one
of the equilibrium points the solution will move into a strange attractor. Looking
at the differential equation for the Lorenz system, it is not obvious at all that the
solution could be chaotic.

In the next chapter, we will discuss the engineering part of the book—the FPGA.

---

[6]If the system has no equilibrium points, you need to rely on intuition to select initial conditions.
As a specific example, refer to Problem 1.8 in the exercises.

Problems

Note that an excellent book on problem solving is "How to Solve It" by Polya [20]. You should obtain a copy of this book and read it thoroughly, it is time well spent!

1.1 Google Scholar http://scholar.google.com is a great tool to find research papers. Using Google Scholar, find Lorenz's original paper from 1963 [6]. Lorenz's paper is very well written, read the paper and determine the physical meaning of the variables x, y and z in Eqs. (1.1)–(1.3). Note that Google Scholar is not yet universally accepted as a scientific database. Hence the reader should be familiar with other tools such as Web of Science and Scopus.

1.2 Consider three dimensional space $\mathbb{R}^3$. Draw all possibilities for a linear system of three equations (in three unknowns x, y, z) such that the system has no solution, unique solution and infinitely many solutions.

1.3 Consider the amplifier system and the system that squares the input function.

$$y_1(t) = S(x)(t) = Kx(t) \tag{1.71}$$

$$y_2(t) = S(x)(t) = x^2(t) \tag{1.72}$$

What is the output of the amplifier if the input is $x(t) = \sin(\omega t)$? What is the frequency of the output? Now, what is the output frequency of the square system when the input is $x(t) = \sin(\omega t)$? Based on the results of this problem, what can you conclude about the output of a nonlinear system to a sinusoidal input of a specific frequency? Can your conclusion be extended to nonlinear dynamical systems?

1.4 Recall the Sprott system from the text with the signum nonlinearity. We wrote the three first-order differential equations as one third order differential equation. Can we, in general, write *any* n first-order differential equations as one nth order differential equation? Prove your answer or give a counter-example. **HINT:** Before you look for a proof, try to write the Lorenz system as one third order differential equation.

1.5 Analogous to the Problem 1.4, can we write *any* nth order differential equation as n first-order differential equations of the form $\dot{\mathbf{x}} = \mathbf{f}(\mathbf{x})$? Prove your answer or give a counter-example.

1.6 Consider the simple harmonic oscillator from the text:

$$m\ddot{x} + kx = 0 \tag{1.73}$$

Can you realize this system physically (on a breadboard, for instance)?

1.7 For each of the following system, determine equilibrium points analytically. Simulate each system near the equilibrium point to determine dynamics. For two dimensional systems, use pplane.

1. Van Der Pol oscillator

$$\dot{x} = \mu \left(x - \frac{x^3}{3} - y \right) \tag{1.74}$$

$$\dot{y} = \frac{x}{\mu} \tag{1.75}$$

$\mu \in \mathbb{R}, \mu \neq 0$ is a parameter indicating the strength of nonlinear damping. Use $\mu = 1.5$ for simulation. What do you observe for $\mu \ll 1$?

2. Rössler System

$$\dot{x} = -y - z \tag{1.76}$$

$$\dot{y} = x + \alpha y \tag{1.77}$$

$$\dot{z} = \beta + z(x - \gamma) \tag{1.78}$$

$\alpha, \beta, \gamma \in \mathbb{R}$ are parameters of the system. Use $\alpha = 0.1, \beta = 0.1, \gamma = 14$ for simulation.

3. Sprott System

$$\dot{x} = -2y \tag{1.79}$$

$$\dot{y} = x + z^2 \tag{1.80}$$

$$\dot{z} = 1 + y - 2z \tag{1.81}$$

4. Chua System

$$\dot{x} = \alpha[y - x - m_1 x - \frac{1}{2}(m_0 - m_1)(|x + 1| - |x - 1|)] \tag{1.82}$$

$$\dot{y} = x - y + z \tag{1.83}$$

$$\dot{z} = -\beta y \tag{1.84}$$

$m_0, m_1, \alpha, \beta \in \mathbb{R}$ are parameters of the system. Use $m_0 = \frac{-8}{7}, m_1 = \frac{-5}{7}, \alpha = 15.6, \beta = 25.58$ for simulation.

1.8 What are the fixed points for the system below?

$$\dot{x} = y \tag{1.85}$$

$$\dot{y} = -x + yz \tag{1.86}$$

$$\dot{z} = 1 - y^2 \tag{1.87}$$

Construct a three dimensional phase plot in MATLAB.

1.9 What is the phase plot for the one dimensional nonlinear differential equation:

$$\dot{x} = \sin(x) \tag{1.88}$$

1.10 In the text, we considered three dimensional systems. Now consider the following four dimensional system proposed by Rössler [21].

$$\dot{x} = -y - z \tag{1.89}$$
$$\dot{y} = x + \alpha y + w \tag{1.90}$$
$$\dot{z} = \beta + xz \tag{1.91}$$
$$\dot{w} = -\gamma z + \delta w \tag{1.92}$$

$\alpha, \beta, \gamma, \delta \in \mathbb{R}$ are parameters. What are the equilibrium points for this system? Simulate the system in MATLAB, try to pick values for α, β, γ, δ for chaos. Contrast the behavior of this system with the Rössler system in Eqs. (1.76)–(1.78). The system above is the first example of a hyperchaotic system.

1.11 In the text, we discussed "robustness" of the Lorenz attractor. A mathematical approach to quantifying "robustness" is to compute the divergence of a vector field **f** defined by the RHS of the Lorenz system:

$$\mathbf{f} \triangleq (-\sigma x + \sigma y)\hat{x} + (-xz + \rho x - y)\hat{y} + (xy - \beta z)\hat{z} \tag{1.93}$$

$\hat{x}, \hat{y}$ and $\hat{z}$ are the unit vectors in the x, y and z directions respectively. Compute $\nabla \cdot \mathbf{f}$. What can you conclude?

Lab 1: Introduction to MATLAB and Simulink

Objective: To compute equilibrium points and numerically investigate behaviour of dynamical systems.

Theory: Refer to the Appendix for a tutorial on MATLAB and Simulink.

Lab Exercises:

1. After working through the Appendix, simulate all systems from Exercise 1.7.
2. Consider the Lotka-Volterra system in Eq. (1.94).

$$\dot{x} = x(\alpha - \beta y)$$
$$\dot{y} = -y(\gamma - \delta x) \tag{1.94}$$

  a. Determine the equilibrium points of this system.
  b. Using pplane, determine all possible qualitatively different phase portraits for this system, as α, β, γ, δ are changed. Note that the Lotka-Volterra system is available in pplane.

3. pplane7 can be used to plot two dimensional phase plots. MATLAB has an equivalent command called quiver. Moreover, in order to plot vector fields for three dimensional systems, investigate the MATLAB command quiver3. Use quiver or quiver3 to plot vector fields for all systems in Exercise 1.7.

References

1. Steingrube S, Timme M, Worgotter F, Manoonpong P (2010) Self-organized adaptation of a simple neural circuit enables complex robot behaviour. Nat Phys 6:224–230
2. Alligood KT, Sauer TD, Yorke JA (1996) Chaos: an introduction to dynamical systems. Springer, New York
3. Chen G, Ueta T (2002) Chaos in circuits and systems. World Scientific, Singapore
4. Van der Pol B (1927) On Relaxation-oscillations, The London, Edinburgh and Dublin Philos Mag J Sci 2(7):978–992
5. Van der Pol B, Van der Mark J (1927) Frequency demultiplication. Nature 120:363–364
6. Lorenz EN (1963) Deterministic nonperiodic flow. J Atmos Sci 20:130–141
7. Matsumoto T (1984) A chaotic attractor from Chua's circuit. IEEE Trans Circuits Syst CAS 31(12):1055–1058
8. Chua LO (2011) Chua's circuit. In: Scholarpedia. http://www.scholarpedia.org/article/Chua_circuit. Accessed 25 Dec 2012
9. Muthuswamy B et al (2009) A Synthetic inductor implementation of Chua's circuit. In: University of California, Berkeley, EECS Technical Reports. http://www.eecs.berkeley.edu/Pubs/TechRpts/2009/EECS-2009-20.html. Accessed 22 Nov 2014
10. Sprott JC (2010) Elegant chaos. Algebraically simple chaotic flows. World Scientific, Singapore
11. Pecora LM, Carroll TL (1990) Synchronization in chaotic systems. Phys Rev Lett 64:821–824
12. Cuomo KM, Oppenheim AV (1993) Circuit implementation of synchronized chaos with applications to communications. Phys Rev Lett 71:65–68
13. A History of Xilinx (2012). In: Funding Universe. http://www.fundinguniverse.com/company-histories/xilinx-inc-history/. Accessed 26 Dec 2012
14. A History of Altera (2012). In: Funding Universe. http://www.fundinguniverse.com/company-histories/altera-corporation-history/. Accessed 26 Dec 2012
15. Top 5 benefits of FPGAs (2012). In: National Instruments Whitepaper. http://www.ni.com/white-paper/6984/en. Accessed 26 Dec 2012
16. Cornell University (2012) Digital Differential Analyzer. In: ECE5760 Homepage. http://people.ece.cornell.edu/land/courses/ece5760/DDA/index.htm. Accessed 26 Dec 2012
17. Strang G (2009) Introduction to linear algebra. Wellesley-Cambridge Press, Massachusetts
18. Lee EA, Varaiya PP (2011) Structure and interpretation of signals and systems, 2nd edn. http://leevaraiya.org/
19. Polking JC (2011) pplane homepage. http://www.math.rice.edu/~dfield/. Accessed 25 Dec 2012
20. Polya G (1957) How to solve it. Doubleday, Gardent City
21. Rössler OE (1979) An equation for hyperchaos. Phys Lett A 71:155–157

Chapter 2
Designing Hardware for FPGAs

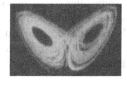

FPGA realization of the Lorenz
butterfly

Abstract In this chapter we will cover many of the basic concepts behind FPGA design. We start with an overview of our hardware platform, go through a quick introduction to the Quartus toolset and then review combinational along with sequential logic. We will conclude with the all important concept of timing closure. Although we cover a particular hardware platform, the material in this chapter can be adopted to understand other FPGA hardware platforms. This chapter, along with Chap. 1, lay the groundwork for the rest of the book. Nevertheless, please understand that majority of this chapter is meant primarily as a review. However, the conceptual material on abstracting the FPGA development flow via Simulink should not be skipped.

2.1 The FPGA Development Flow

In order to design for an FPGA, one needs to intimately understand the design process shown in Fig. 2.1 [1]. The first step in the process is design entry. In other words, you specify design functionality (differential equations) using tools such as Hardware Description Languages (HDLs), schematic entry or using a high level block diagram approach like Simulink. Next we compile the design to identify any syntax errors and then simulate the design to verify functionality. If design specifications are not met, we debug the design entry as necessary in order to meet functional specifications. Once the functional specifications have been met, we should run a

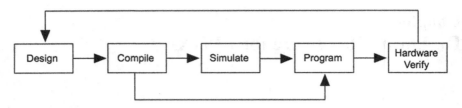

Fig. 2.1 A high-level view of FPGA design flow

timing-intensive simulation, but this topic is beyond the scope of this volume.[1] Once we have confirmed that the design is both functional and satisfies timing, we can download the bitstream onto the FPGA.

The first step in maximizing the capabilities of an FPGA is understanding the underlying architecture, the topic of Sect. 2.2.

2.2 The Architecture of an FPGA

Although the specifics of FPGA architecture vary between each device family (even within the same manufacturer), an FPGA is simply a massively parallel lookup table. Figure 2.2 shows a screen shot from the Quartus chip planner of the FPGA that we will be using in this book, the Cyclone IV.

Note how the device architecture is very repetitive in terms of fundamental structure, i.e.., the FPGA has a two dimensional row and column-based architecture to implement custom logic. Figures 2.3 and 2.4 show just how uniform this structure is.

Let us examine the LE in Fig. 2.4 in some detail [2], since a LE is the basic design unit on an FPGA. Each LE features:

- A four-input look-up table (LUT), which is a function generator that can implement any combinational logic function of four variables
- A programmable register
- A carry chain connection
- A register chain connection
- The ability to drive all types of interconnects: local, row, column, register chain and direct link interconnects
- Support for register packing
- Support for register feedback

An LE can also operate in normal mode or arithmetic mode. Normal mode is suitable for general logic applications and combinational functions. The arithmetic mode is ideal for implementing adders, counters, accumulators and comparators. LEs in

---

[1] We will however discuss the important concept of timing closure.

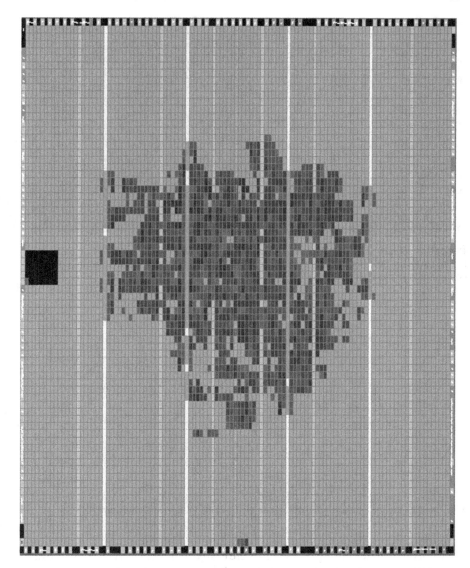

Fig. 2.2 A view of the Cyclone IV from the Chip planner in Quartus. The darker blue indicates filled FPGA regions

arithmetic mode can drive out registered and unregistered versions of the LUT output. Register feedback and register packing are supported when LEs are used in arithmetic mode.

In addition to LEs, Cyclone IV provide Phase Locked Loops (PLLs) for general-purpose clocking [2], as well as support for features such as clock multiplication

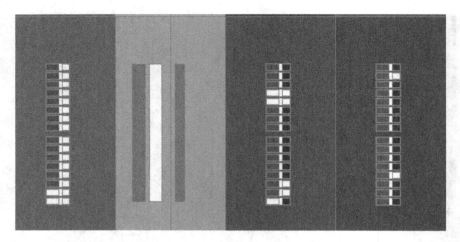

Fig. 2.3 A zoomed in view of the Logic Array Block (LAB) that highlight the 16 Logic Elements (LEs). LABs are the primary design features on the Cyclone FPGA that are grouped into rows and columns across the device [2]

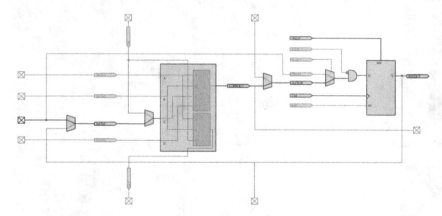

Fig. 2.4 LEs are the smallest units of logic in the Cyclone IV device architecture [2]

(division) and phase shifting. These PLLs are easily configurable via FPGA software tools, this feature will be discussed in the book as necessary. Cyclone IVs also incorporate embedded memory consisting of columns of M9K memory blocks [2]. Each M9K block can implement various types of memory, with our without parity. The M9K blocks are also easy to configure via FPGA software.

We have covered a brief overview of FPGA architecture. Going back to Fig. 2.1, a natural question is: how do we utilize software to realize a design onto an FPGA? The answer is: FPGA manufacturers provide advanced software tools for FPGA hardware design. The toolset that Altera provides is the Quartus suite. However since we are implementing differential equations, we will primarily utilize MATLAB and Simulink. Nevertheless, we first need to choose the appropriate development board.

2.3 An Overview of the Hardware and Software Development Platform

In order to physically realize our differential equations, we will use the DE2-115 board[2] [3] shown in Fig. 2.5 from Terasic Technologies that utilizes a Cyclone IV (EP4CE115F29C7N) FPGA.

One also requires the Quartus toolset from Altera and the MATLAB (along with Simulink) package from Mathworks Corporation. Please contact the respective companies for the appropriate licenses. Note also that you need miscellaneous hardware such as interface cables and oscilloscopes.

FPGA hardware and software platforms evolve rapidly. A decision had to be made on the particular choice of hardware and software. We first chose the DE2-115 because it offered a large functionality-to-cost ratio for our research project(s) and, consequently, we chose the Quartus toolset. This platform was also available at the time when the volume was first published. However, the concepts covered in this volume should be applicable to any appropriate FPGA development platform and toolset(s).

We will now go over the salient features of our development platform, starting with the DE2-115 board.

2.3.1 An Overview of the Terasic DE2-115 Development Board

Since we have already discussed the FPGA in Sect. 2.2, we will discuss board peripherals.

2.3.1.1 FPGA Clocks

Probably the most important component on the FPGA board is the crystal oscillator for the clock circuitry [4]. The DE2-115 board includes one oscillator that produces 50 MHz clock signal. A clock buffer is used to distribute 50 MHz clock signal with low jitter to the FPGA. The board also includes two Subminiature Version A (SMA) connectors which can be used to connect an external clock source to the board or to drive a clock signal out through the SMA connector. In addition, all these clock

---

[2]These are not the only possible development platforms that can be used to realize chaotic dynamics. Please utilize the companion website to obtain information on other development platforms and software tools.

Fig. 2.5 The DE2-115 board [3]

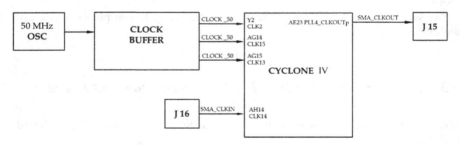

Fig. 2.6 Block Diagram of Clock Distribution on the DE2-115 [4]

inputs are connected to the PLL clock input pins of the FPGA to allow users to use these clocks as a source clock for the PLL circuit [4] (Fig. 2.6).

Since clocks are fundamental to FPGA functionality, pin assignments for clock inputs to FPGA input/output (I/O) pins are listed in Table 2.1.

2.3.1.2 Switches and Light Emitting Diodes (LEDs)

The DE2-115 board provides four push-button switches as shown in Fig. 2.7 [4]. Each of these switches is debounced using a Schmitt Trigger circuit, as indicated in

Table 2.1 Pin assignments for clock inputs

Signal name	FPGA pin no.	Description	I/O standard
CLOCK_50	PIN_Y2	50 MHz clock input	3.3 V
CLOCK2_50	PIN_AG14	50 MHz clock input	3.3 V
CLOCK3_50	PIN_AG15	50 MHz clock input	Depending on JP6
SMA_CLKOUT	PIN_AE23	External (SMA) clock output	Depending on JP6
SMA_CLKIN	PIN_AH14	External (SMA) clock input	3.3 V

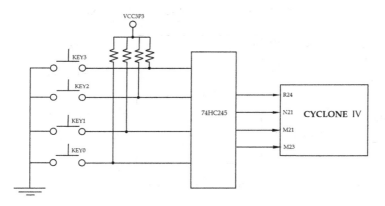

Fig. 2.7 Connections between the push-button and Cyclone IV FPGA [4]

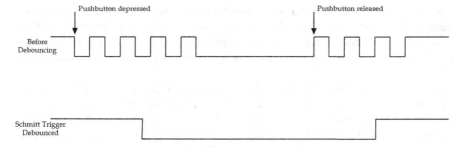

Fig. 2.8 Push button debouncing [4]

Fig. 2.8. The four outputs called KEY0, KEY1, KEY2, and KEY3 of the Schmitt Trigger devices are connected directly to the Cyclone IV E FPGA. Each push-button switch provides a high logic level when it is not pressed, and provides a low logic level when depressed. Since the push-button switches are debounced, they are appropriate for using as clock or reset inputs in a circuit.

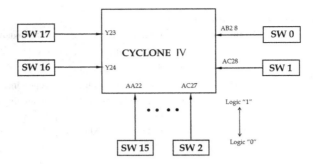

Fig. 2.9 Connections between the slide switches and Cyclone IV FPGA [4]

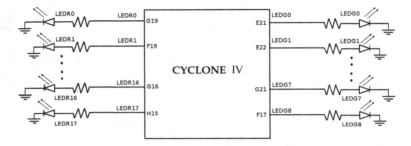

Fig. 2.10 Connections between the LEDs and Cyclone IV FPGA [4]

There are also 18 slide switches on the DE2-115 board (See Fig. 2.9) [4]. These switches are not debounced, and are assumed for use as level-sensitive data inputs to a circuit. Each switch is connected directly to a pin on the Cyclone IV FPGA. When the switch is in the down position (closest to the edge of the board), it provides a low logic level to the FPGA, and when the switch is in the up position it provides a high logic level.

There are 27 user-controllable LEDs on the DE2-115 board [4]. Eighteen red LEDs are situated above the 18 Slide switches, and eight green LEDs are found above the push-button switches (the 9th green LED is in the middle of the 7-segment displays). Each LED is driven directly by a pin on the Cyclone IV FPGA; driving its associated pin to a high logic level turns the LED on, and driving the pin low turns it off. Figure 2.10 shows the connections between LEDs and Cyclone IV FPGA.

2.3.1.3 7-Segment Displays

The DE2-115 board has eight 7-segment displays. These displays are arranged in two pairs and a group of four. As indicated in the schematic in Fig. 2.11, the seven segments (common anode) are connected to pins on the Cyclone IV. The 7-segment displays are actve low.

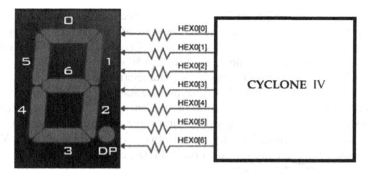

Fig. 2.11 Connections between the 7-segment display HEX0 and the Cyclone IV FPGA [4]

Table 2.2 JP7 settings for different I/O standards

JP7 jumper settings	Supplied voltage to VCCIO5 and VCCIO6 (V)	IO voltage of HSMC connector (JP8) (V)
Short pins 1 and 2	1.5	1.5
Short pins 3 and 4	1.8	1.8
Short pins 5 and 6	2.5	2.5 (Default)
Short pins 7 and 8	3.3	3.3

Table 2.3 JP6 settings for different I/O standards

JP6 jumper settings	Supplied voltage to VCCIO4 (V)	IO voltage of expansion header (JP5) (V)
Short pins 1 and 2	1.5	1.5
Short pins 3 and 4	1.8	1.8
Short pins 5 and 6	2.5	2.5
Short pins 7 and 8	3.3	3.3 (Default)

2.3.1.4 I/O Standards

The I/O voltage levels (standards) for the High Speed Mezzanine Card (HSMC) and the expansion header on the DE2-115 can be set by using JP7 and JP6 respectively [4]. Nevertheless, these jumpers also set the I/O standards for peripherals, for example the clock input in Table 2.1. Hence we list the JP7 and JP6 settings in Tables 2.2 and 2.3.

2.3.2 VHDL Primer and Using the Quartus Toolset

Although we will be primarily utilizing Simulink for a high level functional specification of our chaotic systems, it is important to have an idea of the underlying hardware to which our design synthesizes to. There are many abstraction levels for understanding synthesized hardware. In this book, we will utilize the HDL approach. Specifically, we will use VHDL—one of the two HDLs that have an IEEE standard associated with them [5]. The other IEEE standard HDL is Verilog. We use VHDL in this book because it has better support for parameterized design [6].

Discussing every nuance of VHDL is beyond the scope of this book. Fortunately, many details are abstracted away because of our functional approach to specifying chaotic systems. However, we will still discuss some of the most important ideas behind VHDL in this section. For further study, two very good references on VHDL are Brown and Vranesic that deals with basic VHDL concepts [7] and Chu's book on VHDL for efficient synthesis [6].

Before we begin, an important note: as the name indicates, HDL describes hardware [6]. We are not writing a software program and hence it is essential to approach HDL from the hardware's perspective. The synthesis software should also be treated as a tool to perform transformation and local optimization. It cannot alter the original architecture or convert a poor design into a good one [6]. A rule of thumb: if we as humans cannot understand the underlying hardware functionality via the HDL, the synthesizer will not translate the design into a correct functional specification.

2.3.2.1 VHDL Modules: Entity, Ports and Architecture

Listing 2.1 below is a sample VHDL hardware specification.

Listing 2.1 Combinational logic in VHDL

```
1    -- Lines starting with -- are comments in VHDL.
2    library ieee; -- include ieee library for the use statements
         below
3    use ieee.std_logic_1164.all;
4    use ieee.numeric_std.all;
5
6    entity simpleLogicGates is port (
7        x1,x2,x3 : in std_logic;
8        x : in std_logic_vector(3 downto 0);
9        y : out std_logic_vector(4 downto 0));
10   end simpleLogicGates;
11
12   architecture combinational of simpleLogicGates is
13   begin
14       y(0) <= (x1 and x2) or x3;
15       y(2 downto 1) <= (x1&x(0)) AND (x2&x(1));
16       y(4 downto 3) <= x(3 downto 2);
17   end combinational;
```

The first three statements are IEEE standardized libraries to facilitate synthesis [6]. Line 1 invokes the IEEE library and line 2 allows us to use the predefined datatypes,

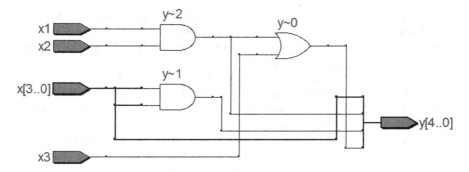

Fig. 2.12 Schematic diagram of our hardware specification, generated using the Quartus RTL viewer. The thicker lines imply multi-bits or a bus

std_logic and std_logic_vector. The numeric_std package enables us to utilize other datatypes such as signed and unsigned.

The entity declaration in line 6 describes the external interface of our circuit [6]. Our design communicates with the outside world via ports. These ports can be input, output or inout (bidirectional). In our example, we have three 1-bit wide input ports (x1, x2, x3) and two multi-bit wide (bus) ports. Note that the std_logic (and std_logic_vector) types support more than '1' and '0'. We will additionally only utilize the high impedance ('Z') support in std_logic, as this is the only other type defined for proper synthesis. A potential issue of utilizing 'Z' is support for tri-state devices in the underlying hardware but the Cyclone IV on the DE2-115 does support these devices. Also note that for interfacing to external physical components, we should not use VHDL types such as integers.[3]

The architecture body specifies the internal operation or organization of the digital logic system [6]. The architecture body consists of concurrent statements that describe combinational logic and process statements that describe sequential logic. We will first give examples of combinational logic design.

2.3.2.2 VHDL Combinational Logic Design and Using Quartus

Going back to our design specification in Sect. 2.3.2.1, we have three concurrent statements that synthesize to the RTL description in Fig. 2.12.

We will now summarize the main steps for using Quartus.

1. The first step in using Quartus is to obtain the software. Although the Quartus web-edition [8] will suffice for this section, we will need the Quartus Subscription Edition [9], ModelSim-Altera [10] and DSP Builder [11] for synthesis, simulation and mathematical functionality specification respectively. Please contact Altera

---

[3]Of course, we are free to choose any type for internal communication between modules. Such flexibility is the purpose of abstraction.

corporation for details on obtaining academic licenses or purchasing software. In this book, we will utilize the 12.0 versions of the toolset, although any version after 12.0 is acceptable. Note that some of the screens and menu actions may be different in newer versions of Quartus.

2. Next we need to download the system CD for our board, the DE2-115, from the board manufacturer (Terasic) website [3]. Although a system CD is part of the board kit, the latest version is available online. The CD has a plethora of useful documentation and reference designs.

3. Now that we have Quartus installed and the system CD, we can start our design by creating a new folder [4] for the combinational logic project.

Please refer to the online video and reference design on the companion website: http://www.harpgroup.org/muthuswamy/ARouteToChaosUsingFPGAs/ReferenceDesigns/volumeI-ExperimentalObservations/chapter2/combinationalLogicDesign/ for completing the simple combinational logic design.

Let us examine a frequently used VHDL construct for combinational logic design, the selected signal assignment, shown in listing 2.2. The selected signal assignment synthesizes to a multiplexer at the RTL level.

Listing 2.2 VHDL selected signal assignment

```
1  with selectBits select
2      output0 <= input0 when "00",
3                 input1 when "01",
4                 input2 when "10",
5                 input3 when others;
```

The careful reader should have noticed that the type of input (and consequently output) cannot be inferred from the VHDL snippet above. The type could be std_logic, std_logic_vector or integers. In fact, now would be a good time to look at the online video for a reference design that realizes an arithmetic logic unit using a mux at the output for selecting between different operations: http://www.harpgroup.org/muthuswamy/ARouteToChaosUsingFPGAs/Reference Designs/volumeI-ExperimentalObservations/chapter2/alu/.

The video should help the reader complete the simple combinational logic design. An RTL view of the ALU is shown in Fig. 2.13.

2.3.2.3 VHDL Parameterization

In this section, we will utilize parameterization functionality of VHDL. We will let the synthesizer infer and connect multiple instances of the same module. Although we will be using a for loop for synthesis, please remember that we are designing hardware, not programming. The Quartus RTL view of the top level from our design is shown in Fig. 2.14.

---

[4] It is not a good idea to include spaces in the project path.

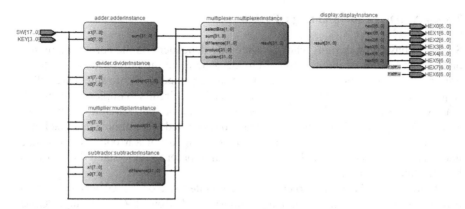

Fig. 2.13 RTL view for the ALU design

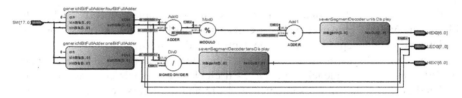

Fig. 2.14 RTL view for the ripple carry adder. Extraction of digits in the sum is done at the top level

The design primarily contains three components:

1. oneBitFullAdder: The oneBitFullAdder simply implements a structural one bit adder. The boolean equations for the sum input and the carry outputs for the ith one bit full adder are shown in Eqs. (2.1) and (2.2) respectively.

$$s_i = x_i \oplus y_i \oplus c_i \tag{2.1}$$
$$c_{i+1} = x_i y_i + c_i(x_i + y_i) \tag{2.2}$$

2. genericNBitFullAdder: In order to realize the genericNBitFullAdder, we will connect n one bit full adders in a ripple carry structure, as shown in listing C.1.
3. sevenSegmentDecoder: The seven segment decoder is a standard decimal to hex decoder module. The VHDL description is shown in listing C.2.

The top level realization of the generic ripply carry adder is shown in listing C.3. In order to completely understand the design, please look at the online video for the genericNBitFullAdder: http://www.harpgroup.org/muthuswamy/ARouteToChaosU singFPGAs/ReferenceDesigns/volumeI-ExperimentalObservations/chapter2/ripple CarryAdder/.

2.3.2.4 VHDL Sequential Logic Design

So far we have seen designs whose output only dependent on the current input, not on the past history of inputs. Digital circuits where the input depends on both the present and past history of inputs (or the entire sequence of input values) are called sequential logic circuits [6]. We need sequential logic circuits to implement memory via registers (flip-flops). We store the system's state or state variables in memory and hence sequential logic circuits can also be specified using finite state machines (or state machines). Figure 2.15 shows a block diagram of a Moore (Mealy) state machine. We will examine a 24-h clock design in order to understand the concepts behind state machines.

The synthesized Quartus project and a 20-minute video on the design can be obtained from: http://www.harpgroup.org/muthuswamy/ARouteToChaosUsingFPGAs/ReferenceDesigns/volumeI-ExperimentalObservations/chapter2/twentyForHourClock/.

Considering Fig. 2.15, one can infer that we should be able to specify each of the blocks via VHDL. However before we discuss the VHDL realization, we need to understand the concept of a globally synchronous design [6].

A large digital system should consist of subsystems that operate at different rates or different clock frequencies. In our 24-h clock, the seconds counter should be updated at 1 Hz, the minutes counter at $\frac{1}{60}$ Hz and the hours counter at $\frac{1}{3600}$ Hz. There are primarily two approaches for clocking, shown in Figs. 2.16 and 2.17.

There are two main problems with the approach in Fig. 2.16. First, the system is no longer synchronous because if the subsystems interact, then the timing analysis becomes very involved. Second problem is the placement and routing of multiple clock signals. Since a clock signal needs a special driver and distribution network, having derived clock signals makes this process more difficult.

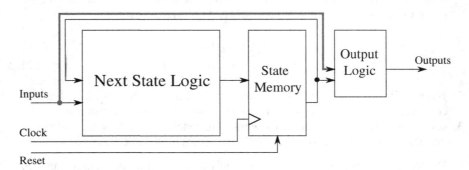

Fig. 2.15 Generic block diagram for a finite state machine. State machines are composed of next state logic (a combinational function of inputs and synchronous current state). The output function combinational logic function can be a function of current state only (Moore machine) or current state and input (Mealy). There is a single clock to ensure the design is fully synchronous. All finite state machines must have a well defined global reset

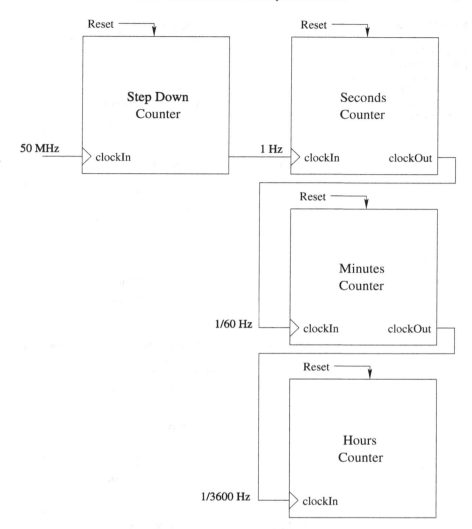

Fig. 2.16 System with multiple clock frequencies. The global clock is stepped down to a 1 Hz clock and each module subsequently outputs clocks with different frequencies

In contrast, the low rate single-clock enable pulse design shown in Fig. 2.17 is the preferred approach since the different subsystems are driven with the same clock frequency.

The specification of the seconds counter, along with a single pulse generator is in listing C.4 (obtained from the online Quartus project). The single pulse generator is a state machine that helps us implement the scheme shown in Fig. 2.17. The state machine generates a pulse that is exactly one clock cycle long, everytime the seconds counter overflows.

Fig. 2.17 System with a
single synchronous clock.
Each module uses a single
pulse generator that has an
enable pulse that is exactly
20 ns wide. This pulse acts as
a synchronous trigger input
for the subsequent module

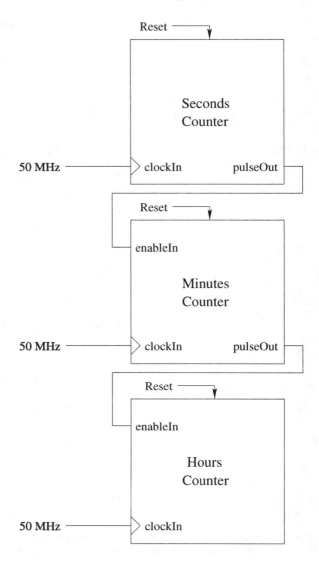

One way to visualize an FSM is using State Machine Diagrams. The state transition
diagram for the single pulse generator is shown in Fig. 2.18.

Once you download the online design to the DE2-115 board, you will notice that
the base design clock is counting much faster than the usual 1 Hz frequency for a
seconds counter. Exercise 2.1 asks you to modify the design so we have the 1 Hz
frequency for the seconds counter.

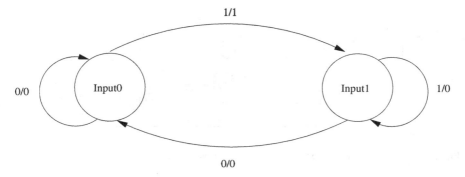

Fig. 2.18 A state transition diagram for the Mealy FSM realization of the single pulse generator. The circles represent states and the arcs represent transitions between states. Input(s)/Output(s) are specified on the arcs. It is assumed that transitions take place only on clock edges and hence synchronous behaviour is implied. Hence, the output is logic 1 only when we transition from the Input0 to Input1 state, in other words, the output is high only for one clock pulse. A Moore FSM is specified in a similar manner, except the outputs are given in the states. How would you specify the single pulse generator as a Moore FSM? *Hint* you need at least one more state

You should also understand that the design has a well defined reset state. Since the DE2-115 keys are debounced in hardware, we can utilize them as reset inputs without a debounce circuit. Exercise 2.4 considers debouncer design.

Going back to the issue of multiple clock frequencies in a design, a natural question is: can we always use a single clock frequency for every design? The answer is: no. This situation is very common when interfacing to external devices. For example, if our design were to interface with external memory (like SDRAM) then we will most likely have our FPGA design running at one frequency while the external SDRAM's clock is at a different frequency. Nevertheless we should clearly separate the different clock domains in our design. If we do, then we can utilize powerful timing closure tools provided by FPGA manufacturers to ensure that our design meets timing requirements. We will discuss timing closure in Sect. 2.4. In fact the next section shows an example design where we do interface to external hardware on the DE1 board.

2.3.3 Audio Codec Interfacing

We need ADC and DAC converters to interface external signals to/from the FPGA respectively. On the DE2 board, there is a Wolfson WM8731 [12] audio coder/decoder (codec) that has on-board ADC and DAC. This section shows how we can interface to these peripherals.

Figure 2.19 shows a top-level block diagram of our design. A discussion of each block follows [13].

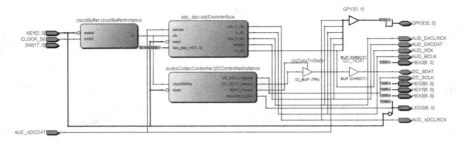

Fig. 2.19 A screenshot from the Quartus RTL view of our i^2c interface

The clock buffer uses two PLL modules: one to buffer the 50 MHz global clock and the other to provide a 18.42105 MHz clock (from the 27 MHz clock) for sending data to/from the codec.

The i^2c interface initializes the audio codec by sending a series of 10 data packets. In accordance with the i^2c protocol, each data packet is 24 bits long, consisting of: the codec address, the control register and the control register settings. The 10 packets were retrieved from ROM, and sent over the two-wire interface via an FSM.

The i^2c clock signal was a 50 KHz clock signal, generated from the 50 MHz PLL buffered FPGA clock using a counter. Functionally, the i^2c clock lagged the FSM clock by a half-clock cycle. This allowed the FSM to update each bit on the rising edge of the clock but the bit was still correctly interpreted by the codec.

Hence this design utilizes different clock frequencies between modules and this cannot be avoided due to the fact that we are interfacing to an external device. Also, the i^2c protocol limits the maximum frequency of the clock to be 3.4 MHz, depending on the mode [14]. Therefore, there is no possibility of clocking the i^2c initialization module at the global clock frequency of 50 MHz.

The ADC_DAC controller module's primary function is to place data into and read data from the A/D and D/A registers respectively. First, this controller generates two clock signals:

1. A bit clock with a frequency of 3.07 MHz to clock I/O bits from the codec.
2. A frame clock with a frequency of 192 KHz to signify either the left or right channel.

The synthesized Quartus project for our design can be found online:
http://www.harpgroup.org/muthuswamy/ARouteToChaosUsingFPGAs/Reference
Designs/volumeI-ExperimentalObservations/chapter2/DE2i2cInterface/.

The State Machine Viewer in Quartus can be used to examine the state transition diagram of our FSM. We can access the state machine viewer using Analysis and Synthesis→Netlist Viewers→State Machine Viewer.

An important point that we need to remember is the voltage range of the ADA are ±2 V. Hence any digital design's I/O must confirm to this range of voltage. Also, the I/O range assumes there is no loading of the codec input and output.

Fig. 2.20 The experimental setup used in this book. We are not using the *left-channel* in this experiment

Figure 2.20 shows the setup that we used throughout the book. We have a stereo-to-banana cable that is used for interfacing to scope probes. Figure 2.21 shows the results of the classic loop-back test: the input from the ADC is directly to DAC output (the online i^2c reference design utilizes loopback).

Nevertheless, only the phase delay is apparent in Fig. 2.21. Effects of sampling are not readily apparent with a sine wave. Figure 2.22 shows the result of the loop-back rest with a square wave.

Although we could implement the nonlinear functions and the numerical integration method in VHDL,[5] it is much easier to utilize Simulink for an abstract specification of the mathematics. Section C.5 highlights the conceptual steps in using DSP Builder, the Simulink library developed by Altera. You should go through that section before continuing on with the rest of the this chapter.

---

[5]That is, one could use the MegaWizard in Quartus and avoid DSP Builder. However, the DSP builder approach is more visual and this is the approach that we will use in this book. If you don't have access to DSP Builder, then you can utilize the approach using the MegaWizard. You can discuss questions related to this approach in the online forums available on the book's companion website. Note however that we will use the MegaWizard for implementing some of the functionality, such as bifurcations in Chap. 4.

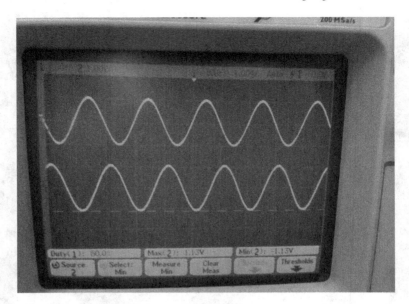

Fig. 2.21 Sine wave loopback, input frequency is 500 Hz. Notice the large phase delay at the output

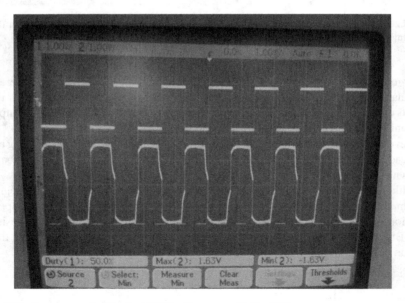

Fig. 2.22 Square wave loopback at 700 Hz. Compare with Fig. 2.21

2.4 Timing Closure

In this section we will discuss the concept of timing closure [5, 15] and look at an example of timing closure for the 24 h clock design from Sect. 2.3.2.4. Although reference designs on the companion website are closed with respect to timing, we will leave the advanced timing closure principles to volume II.

At the start of FPGA technology in the 1980 s, signal propagation delay in logic gates was the main contributor to circuit delay, while wire delay was negligible [15]. Hence cell placement and wire routing did not noticeably affect the final FPGA design. However, starting the in late 990 s, the advent of high density FPGAs and the consequent increase in the size of the final FPGA design implied that there was a need for automated timing closure tools.

Simply put, timing closure is the process by which we ensure the final placed and routed design on the FPGA satisfies timing requirements: setup and hold times for all flip-flops in our design are not violated. Timing closure tools such as TimeQuest (included in Quartus) from Altera adjust propagation delays in the final netlist such that the primary goal of setup and hold time constraints are satisfied. Recall from our basic logic courses that setup (long-path) constraints specify the amount of time a data input signal should be steady before the clock edge for each storage element. Hold time constraints specify the amount of time a data input signal should be stable after the clock edge.

Setup time constraints ensure that no signal transition occurs too late. Initial phases of timing closure focus on these types of constraints, as formulated[6] in Eq. (2.3).

$$t_{\text{clockPeriod}} > t_{\text{combinationalDelay}} + t_{\text{setupTime}} + t_{\text{skew}} \qquad (2.3)$$

In Eq. (2.3):

1. $t_{\text{combinationalDelay}}$ is the worst-case combinational logic delay
2. $t_{\text{setupTime}}$ is the setup time of the receiving flip-flop
3. t_{skew} is the clock skew—the maximum time difference between flip-flop clock edges

Setup constraints are usually performed as part of static timing analysis, which defines timing slack as the difference between required arrival time and actual arrival time, as shown in Eq. (2.4). Positive slack means timing requirements have been met.

$$\text{Timing Slack} = \text{Required Arrival Time} - \text{Actual Arrival Time} \qquad (2.4)$$

Hold time constraints ensure that signal transitions do not occur too early [15]. Hold violations can occur when a signal path is too short, allowing a receiving flip-flop to capture the signal at the current cycle instead of the next cycle. Thus the hold time constraint is formulated as in Eq. (2.5).

---

[6]Some authors define $\geq$ instead of $>$ in Eq. (2.3). We have considered the worst-case scenario and thus use $>$.

$$t_{combinationalDelay} > t_{holdTime} + t_{skew} \hspace{3cm} (2.5)$$

Note that clock skew usually affects hold time constraints than setup time constraint. Thus hold time constraints are typically enforced after placing and routing the clock network [15].

In order to experimentally understand these concepts, examine the SDC specification for timing constraints in Sect. C.5. TimeQuest uses the SDC file to close timing. For further instructions, please look at the video on the companion website: http://www.harpgroup.org/muthuswamy/ARouteToChaosUsingFPGAs/ReferenceD esigns/volumeI-ExperimentalObservations/chapter2/twentyFourHourClock/.

2.5 Conclusions

Below is a summary of the main concepts in this chapter:

1. The FPGA is an ideal platform for implementing discrete specifications of non-linear differential equations because of the massively parallel architecture and variable (user-specified) data and address bus widths. Nevertheless, properly utilizing an FPGA requires the user to have a sound knowledge of basic digital logic principles.
2. In the case of sequential logic, one must aim for a globally synchronous design.
3. For implementing abstract mathematical concepts, we will use DSP Builder Advanced Blockset from Altera.
4. Timing closure is the process of satisfying timing constraints by informing the timing closure tool as to how the design should operate (with respect to timing parameters). The industry standard SDC file is used for specifying timing parameters to TimeQuest, the timing closure tool included with Quartus.

This chapter involved a lot of ideas and hopefully most of them were review of digital logic design concepts. In Chap. 3, we will combine digital logic design concepts and the ideas of DSP builder from this chapter to realize some classic chaotic systems on FPGAs via Simulink.

Problems

2.1 Modify the 24-h clock design from Sect. 2.3.2.4 to accurately count seconds, minutes and hours.

2.2 Instantiate the D flip-flops for synchronous reset from the 24-h clock design in VHDL as opposed to a component-based specification.

2.3 Design a finite state machine whose output is a 1 iff there are two consecutive 1 s in the input stream.

1. Design a Moore FSM for this problem.
2. Design a Mealy FSM for this problem.

2.4 In this problem, we will consider debouncer design [16]. The goal is to design a circuit that would compensate for the mechanical bounces in switch contacts. This circuit is necessary because consider our 50 MHz system clock with 20 ns period. Say a mechanical bounce lasts for 1 ms.

1. How many system clock cycles is one mechanical bounce?
2. Let us say we decide to have a timing-based solution: we declare an input change after signal has been stable for at least 5 ms. Design a system that incorporates a finite state machine and timer to accomplish this task.

Test your realization by using the mechanical switches on the DE2 board.

2.5 Design a finite state state machine that returns the remainder when an arbitrary size integer is divided by 5. One way to test your design on the DE2 board is: use two keys as 0 and 1 inputs. You can use an other key to send a "display remainder" signal to your design. Obviously, the fourth key is global reset.

2.6 The concept of recursion is central to computer programming. Consider listing 2.3 that recursively defines the factorial function:

Listing 2.3 Recursive specification of factorial function in MATLAB

```
1   function y = myfact(number)
2   %myfact Recursive realization of factorial function:
3   % n! = n*(n-1)*(n-2)...1
4       if number == 0
5           y=1;
6       else
7           y=number*myfact(number-1);
8       end
9   end
```

A natural question to ask would be: are there recursive structures in hardware? The answer is yes and a classic example is specifying an m-bit 2^n-to-1 mux (m-bits is the input/output bus width with n-select bits) using 2-1 muxes. Using Fig. 2.23 as an example, design and realize on the DE2 board a recursive multiplexer specification.

The elegant solution in Fig. 2.23 was proposed by Jake Nagel in the EE2900 (Combinational Logic Design) course at the Milwaukee School of Engineering in the Winter 2012–2013 quarter.

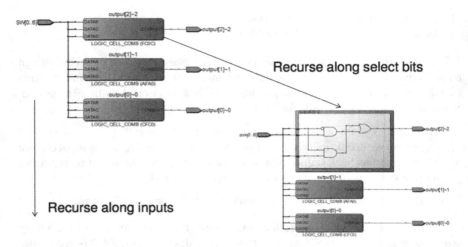

Fig. 2.23 Recursively specifying a 3-bit 2-to-1 mux, technology map viewer from Quartus

Lab 2: Introduction to Altera FPGA Tools

Objective: DE2 LCD interface.

Theory: We first need to thoroughly understand the LCD communication protocol. The DE2-115 user's manual [4] should be our starting point. The display controller is the HD44780 and a data sheet is available on the system CD that accompanies the DE2 board. Nevertheless, you can simply search online and get the latest version of the data sheet.

Lab Exercise:

After going through the LCD documentation, design an FSM to display the words "Hello" on the first line and "World" on the second line of the LCD display. This lab should be a very good review of digital logic concepts, so please take your time to complete the design before looking at the online solution video.

References

1. Altera Corporation (2008) My First FPGA Design Tutorial. In: Altera Corporation Online Tutorials, http://www.altera.com/literature/tt/tt_my_first_fpga.pdf. Accessed 22 Mar 2013
2. Altera Corporation (2013) Cyclone IV Handbook. In: Altera Corporation Online Datasheets, http://www.altera.com/literature/hb/cyclone-iv/cyclone4-handbook.pdf. Accessed 19 Apr 2013
3. Terasic (2013) Altera DE2-115 Development and Education Board. In: Terasic Corporation Online Cyclone Main Boards, http://www.terasic.com.tw/cgi-bin/page/archive.pl?Language=English&No=502. Accessed 21 Apr 2013

4. Terasic (2013) Altera DE2-115 Board User's Manual on the System CD. In: Terasic Corporation Online Cyclone Main Boards, http://www.terasic.com.tw/cgi-bin/page/archive.pl?Language= English&No=502. Accessed 21 Apr 2013

5. Simpson P (2010) FPGA design—best practices for team-based design. Springer, New York

6. Chu PP (2006) RTL hardware design using VHDL—coding for efficiency, portability and scalability. Wiley-Interscience, New Jersey

7. Brown S, Vranesic Z (2008) Fundamentals of digital logic design with VHDL, 3rd edn. McGraw-Hill, New York

8. Altera Corporation (2013) Quartus Web Edition, http://www.altera.com/products/software/ quartus-ii/web-edition/qts-we-index.html. Accessed 7 May 2013

9. Altera Corporation (2013) Quartus Subscription Edition, http://www.altera.com/products/ software/quartus-ii/subscription-edition/qts-se-index.html. Accessed 7 May 2013

10. Altera Corporation (2013) ModelSim-Altera Edition, http://www.altera.com/products/ software/quartus-ii/modelsim/qts-modelsim-index.html. Accessed 7 May 2013

11. Altera Corporation (2013) DSP Builder, http://www.altera.com/products/software/products/ dsp/dsp-builder.html. Accessed 7 May 2013

12. Wolfson WM8731 datasheet, http://www.wolfsonmicro.com/products/audio_hubs/WM8731/. Accessed 4 Oct 2013

13. Stapleton C (2011) Neuron Project EE2902 Spring 2011 Final Project Report

14. NXP Semiconductors (2012) UM10204 i^2C-bus specification and user manual. http://www. nxp.com/documents/user_manual/UM10204.pdf. Accessed 30 Sep 2013

15. Kahng A et al (2011) VLSI physical design: from graph partitioning to timing closure. Springer, New York

16. Brigham-Young University (2013) ECEn 224 Debouncing a Switch - A Design Example, Available via DIALOG. http://ece224web.groups.et.byu.net/lectures/20%20DEBOUNCE. pdf. Accessed 12 Oct 2013

Chapter 3
Chaotic ODEs: FPGA Examples

Abstract In this chapter, we will focus on realizing chaotic systems on an FPGA. We will first show a simple numerical method for specifying chaotic systems on the FPGA and then realize the Lorenz system. We will then illustrate the complete FPGA design process of functional simulation, in-system debugging and physical implementation. In order to illustrate the robustness of FPGAs, we will conclude this chapter by realizing a chaotic system with a hyperbolic tangent nonlinearity.

3.1 Euler's Method

Recall from Chap. 1 that we studied chaos in continuous-time differential equations. We also learned that an FPGA can be used to realize a sampled and discretized version of the differential equation (recall Fig. 1.9). In this section, we will understand that the block diagram in Fig. 1.9 is how we realize the forward-Euler's method [11] on an FPGA.

Consider Eqs. (3.1)–(3.2).

$$\dot{x}_1 = f_1(x_1, \ldots, x_n) \qquad (3.1)$$

$$\vdots$$

$$\dot{x}_n = f_n(x_1, \ldots, x_n) \qquad (3.2)$$

© Springer International Publishing Switzerland 2015
B. Muthuswamy and S. Banerjee, *A Route to Chaos Using FPGAs*, Emergence, Complexity and Computation 16, DOI 10.1007/978-3-319-18105-9_3

We can use first principles [11] and rewrite Eqs. (3.1)–(3.2) as Eqs. (3.3)–(3.4).

$$x_1(t + \delta t) = x_1(t) + f_1(x_1(t), \ldots, x_n(t))\Delta t \qquad (3.3)$$

$$\vdots$$

$$x_n(t + \delta t) = x_n(t) + f_n(x_1(t), \ldots, x_n(t))\Delta t \qquad (3.4)$$

When implementing Eqs. (3.3)–(3.4) on an FPGA, we can define:

$$xNNew \stackrel{\triangle}{=} x_n(t) + f_n(x_1(t), \ldots, x_n(t))\Delta t \qquad (3.5)$$

$$xN \stackrel{\triangle}{=} x_n(t + \delta t) \qquad (3.6)$$

In Eqs. (3.5) and (3.6), $xNNew$ and xN are VHDL signals with $N = 1, 2, \ldots, n$. δt is the clock period for the D flip-flop that results in a synchronous xN. Δt is the step-size in Euler's method. Listing D.1 shows one possible VHDL specification of Euler's method for the Lorenz system. However for consistency (with material from Chap. 1) and clarity (the Lorenz system is three dimensional), we can simply use x, y, z instead of $x1$, $x2$, $x3$. Thus the VHDL design that we will use is shown in listing D.2.

In order to complete listing D.2, we have to specify the various nonlinearities and scale the nonlinearities by Δt. For these purposes, we will utilize DSP builder, along with a VHDL specification of Euler's method. This topic is the subject of Sect. 3.2. But a detailed mathematical discussion of the numerical methods for approximating continuous-time dynamical systems is beyond the course of this book. This will be discussed in more detail in volume II. Simply put a "large enough" step size in Euler's (or any other numerical method) [9] could introduce anomalies in the numerical simulation.

3.2 Specifying Chaotic Systems for FPGAs Using DSP Builder

3.2.1 The Lorenz System

Let us first simulate the Lorenz system in Simulink but using the discrete-time integrator block. Figure 3.1 shows the discrete Euler's method specification of the Lorenz system. We have scaled the state variable outputs to match the output voltage range of the audio codec. $x(t)$ from the simulation plotted via MATLAB is shown in Fig. 3.2. Code for plotting the result is in listing D.3.

Now that we have verified the functionality of a discrete implementation of the Lorenz system, we need to implement the nonlinearities using DSP builder. The procedure incorporates the ideas from Sect. C.5 and an online-video of the implementation can be found on the companion website: http://www.harpgroup.org/muthuswa my/ARouteToChaosUsingFPGAs/ReferenceDesigns/volumeI-ExperimentalObser vations/chapter3/lorenzSystem/.

Discrete-time FIXED step simulation of Lorenz equation using forward Euler.
In order to simulate using physical DE2 realization parameters,
we need the simulation step size to be 64/50e6 = 1.28 us. K needs to be 1/1024
and Ts=-1 for all discrete-time integrator blocks. However, this obviously requires a lot of memory. Therefore, we will use K=1 and reduce
sampling time to 1e-3.

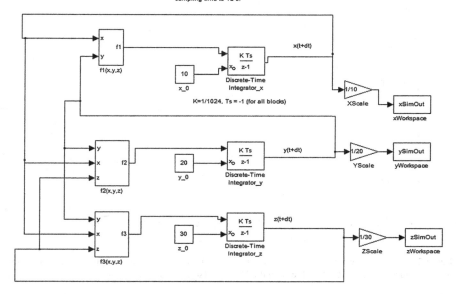

Fig. 3.1 Simulink Euler's method realization of the Lorenz system. We use $K = 1$, $Ts = -1$ ($\Delta t = \frac{1}{1000}$) for each discrete time integrator block and use a sampling time of 1e-3 to conserve memory. The physical implementation uses $\Delta t = 2^{-10}$ and a sampling time of 1.28 μs. The discrete-time integrator block can be found in the Simulink libraries, under Discrete. You can download Simulink design from the companion website: http://www.harpgroup.org/muthuswamy/ARouteToChaosUsingFPGAs/ReferenceDesigns/volumeI-ExperimentalObservations/chapter3/lorenzSystem/

We can now complete the VHDL in listing D.2, as shown in listing D.3. An online video that shows how to incorporate this design with the i^2c interface from Sect. 2.3.3 is on the companion website: http://www.harpgroup.org/muthuswamy/ARouteToChaosUsingFPGAs/ReferenceDesigns/volumeI-ExperimentalObservations/chapter3/lorenzSystem/.

Figure 3.3 shows $x(t)$ and also the FFT of the signal. If you have headphones, you should plug them into the line out port and listen to the sounds of the stereo Lorenz chaotic attractor!

One potential issue with the discrete realization is that we have not really discussed how the our sampling frequency affects the chaotic signal frequency content. We may need to adjust either the sampling frequency or the fixed-point representation. In the case of the Lorenz system, consider Fig. 3.4.

In order to refine[1] $z(t)$ in Fig. 3.4, we changed the fixed-point representation and this is actually reflected in the online reference design.

Figure 3.5 shows the result. Figures 3.6, 3.7 and 3.8 show the various phase plots.

---

[1]One way to predict the effect on $z(t)$ in Fig. 3.4 is to use functional simulation, the subject of Sect. 3.3.

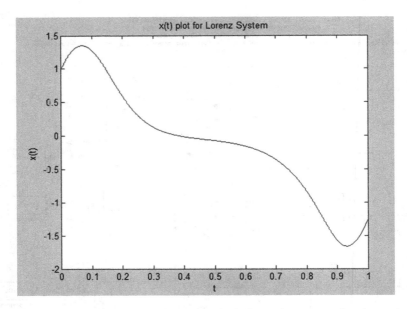

Fig. 3.2 $x(t)$ from the Lorenz system. Notice the plot indicates that our sampling time and range of $x(t)$ values should be compatible with the DE2 Wolfson audio codec parameters

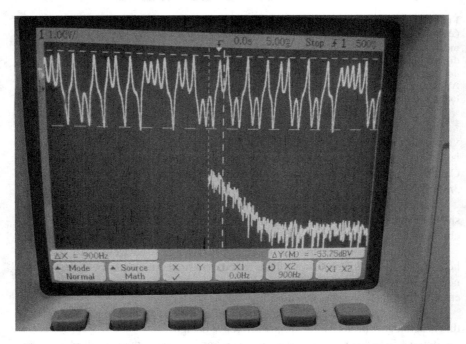

Fig. 3.3 $z(t)$ from the Lorenz attractor realization on the FPGA, along with an FFT of the signal

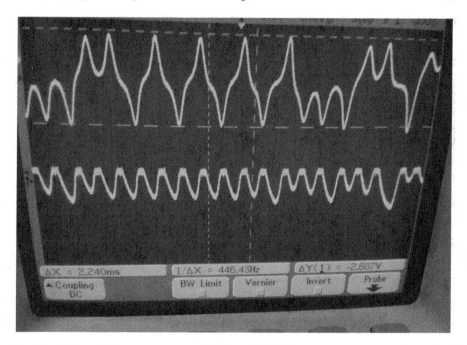

Fig. 3.4 $x(t)$ from the Lorenz attractor realization on the FPGA, compared to $z(t)$. Notice the large variations in $z(t)$ values are not being adequately captured by our digital system

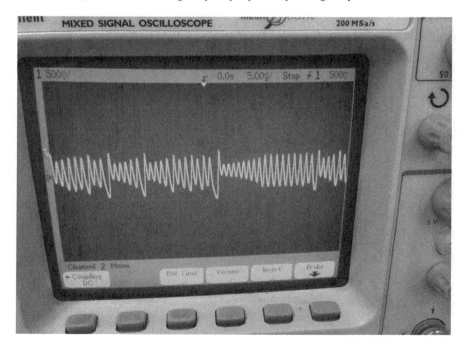

Fig. 3.5 We changed the fixed-point representation to refine $z(t)$

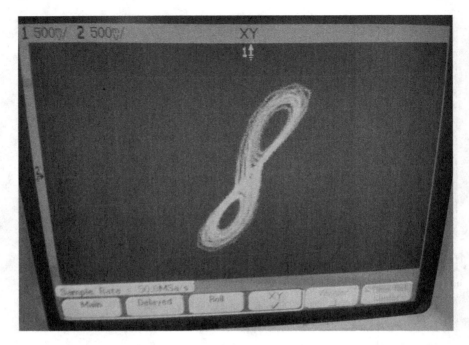

Fig. 3.6 $y(t)$ versus $x(t)$ Phase plot of the Lorenz attractor, the *left* (*right*) audio channel is on input X or channel 1 (input Y or channel 2) of the scope. Screen intensity has been set to 50 %

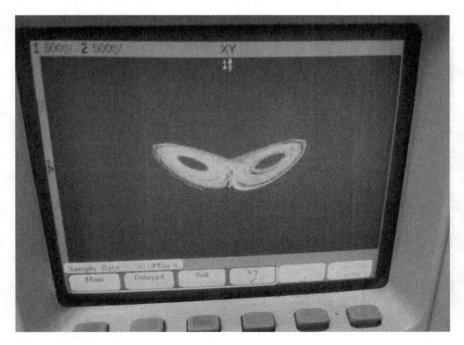

Fig. 3.7 $z(t)$ versus $x(t)$—the classic Lorenz butterfly

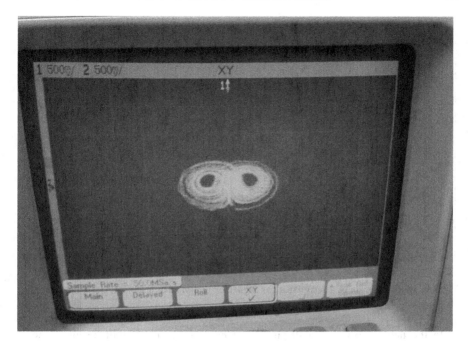

Fig. 3.8 $z(t)$ versus $y(t)$

3.3 Introduction to Functional Simulation and In-System Debugging

After we complete our design in Simulink, it would be prudent to check if the design is functionally correct. That is, does the hardware perform as it is supposed to? In the case of chaotic systems, a functional simulation will also tell us if our design functionally reproduces chaotic behavior. One could also perform a timing-intensive simulation that will also account for delays in the physical FPGA [4]. Nevertheless, a functional simulation is more than enough for us to check the effects of sampling rate. Moreover, timing simulations take a lot longer time to run and we will not be pursuing them further in this book.

There is another important difference between a functional simulation and checking our system behaviour in Simulink. Functional simulation can tell us how the signals propagate within our hardware system. Hence a functional simulation will test the fundamental correctness of our digital circuit. We will use the industry standard ModelSim simulator for functional simulation.

As a natural followup to functional simulation, we can utilize a tool called Signal-Tap, to debug the design as it executes on the FPGA. The "SignalTap Core" is a logic analyzer provided by Altera that is compiled with your design. A block diagram view of a system that incorporates SignalTap is shown in Fig. 3.9. In-system debugging using SignalTap is discussion in Sect. 3.5.

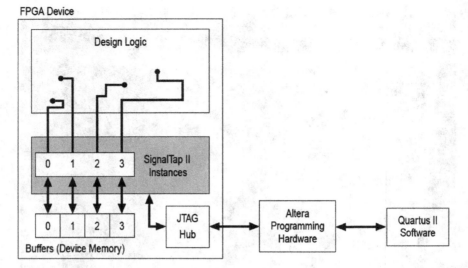

Fig. 3.9 Block diagram of a design that has SignalTap compiled as a separate design partition [2]

In the next section, we will simulate Chen's system. We will discuss general concepts behind simulation and then highlight the main steps as to how we can perform the simulation in ModelSim via the online video. The discussion of ModelSim will be followed by the SignalTap section.

3.4 Functional Simulation of Chaotic Systems

Equations (3.7)–(3.9) are the system equations for Chen's system [7].

$$\dot{x} = a(y - x) \tag{3.7}$$

$$\dot{y} = (c - a)x - xz + cy \tag{3.8}$$

$$\dot{z} = xy - bz \tag{3.9}$$

Parameters are $a = 35, b = 3, c = 28$. Initial conditions are (10, 20, 30).

First, we will perform a discrete Euler simulation and realize this system in DSP Builder. The reference design is placed online: http://www.harpgroup.org/muthuswamy/ARouteToChaosUsingFPGAs/ReferenceDesigns/volumeI-Experime ntalObservations/chapter3/. We have not prepared a video since obviously the steps are the same as in Sect. 3.2.1. Also note that we have not included ModelSim simulation results since the amount of data can be hundreds of megabytes (depends on simulation time length, number of waveforms etc.).

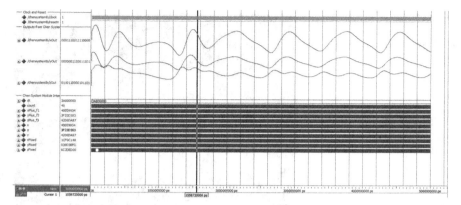

Fig. 3.10 1 ms functional simulation of the Chen system, performed in Modelsim. Notice how ModelSim can interpret the x, y, z state variables from our Chen system as an analog waveform

Fig. 3.11 y(t) versus x(t) for Chen's system. 0.5 V/div for both scales on the digital scope

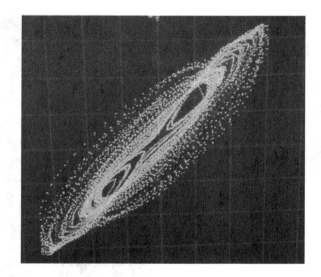

In order to perform functional simulation, we first need to design a VHDL file called as a "test bench", that mimics a physical lab bench [4]. The test bench for the Chen system is shown in listing D.6.

An online video that illustrates how to perform ModelSim simulation is available on the companion website: http://www.harpgroup.org/muthuswamy/ARouteToCha osUsingFPGAs/ReferenceDesigns/volumeI-ExperimentalObservations/chapter3/c henSystem. Once we know the script window commands appropriate to our design, they can be placed in a batch file with a .do extension. This file is also available online and is also shown in listing D.7. In order to use this file, you can simply type: "do FILENAME.do" in the transcript window command line as soon as you start

Fig. 3.12 z(t) versus y(t) for Chen's system. X-axis scale is 0.5 V/div, Y-axis scale is 0.2 V/div

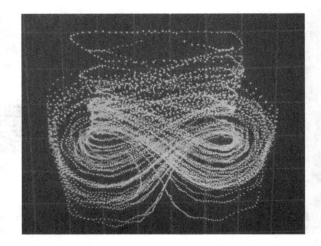

Fig. 3.13 x(t) versus z(t) for Chen's system. X-axis scale is 0.2 V/div, Y-axis scale is 0.5 V/div

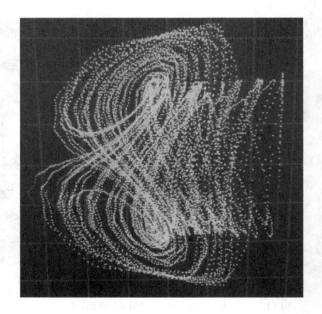

ModelSim (assuming you have already created a default work library with the vlib command).

Figure 3.10 show the result of the ModelSim simulation. Figures 3.11, 3.12 and 3.13 show the phase plots from the physical realization.

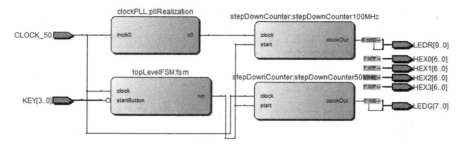

Fig. 3.14 The top level RTL view of the simple SignalTap project

3.5 Debugging Using SignalTap

Before we debug the Chen system in SignalTap, let us go over some general ideas, using a simple design.

3.5.1 General Concepts—An Illustration Using a Simple Example

The design to be analyzed configures a Cyclone II PLL via the MegaIPWizard to step up the 50 MHz board clock to 100 MHz. We use a counter to step down the 50 MHz clock to 1 Hz and output the slow clock to a green LED on the DE2. We use a different instance of the 1 Hz counter, with an input clock of 100 MHz. We output this clock to a red LED on the board. Visually, the red LED will appear to flash twice as fast as the green LED. A top-level FSM waits for the user to press KEY(0) on the DE2 board before running the design. Figure 3.14 shows a top-level RTL view of the project. To avoid confusion, we have not included the SignalTap core in Fig. 3.14.

There are many ways to start[2] SignalTap. The simplest is to add a SignalTap (.stp) file to our project. We can then configure SignalTap, compile SignalTap with our design and download the resulting .sof file to the DE2. Note that every time we change settings in SignalTap, we will have to recompile the entire project, if we do not have Incremental Compilation enabled in Quartus. Incremental Compilation is beyond the scope of this book.

To add the SignalTap file to the project, use **File→New** and select the **SignalTap II Logic Analyzer File** (under Verification/Debugging Files). The SignalTap window in Fig. 3.15 should appear, the different "sub-windows" have been labeled. We will use the "sub-windows" to understand the three primary concepts behind SignalTap:

---

[2]Before using SignalTap, you may need to enable the TalkBack feature in Quartus under **Tools→Options→Internet Connectivity**.

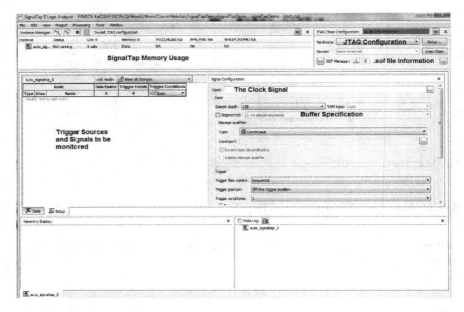

Fig. 3.15 SignalTap main window

the clock signal, trigger and buffer. Please refer to Fig. 3.15 as you read the three ideas below.

1. The Clock Signal provides the sampling clock source for the design, we cannot display this clock in SignalTap. It is implied that SignalTap is always sampling at this clock frequency, data is captured when one or more trigger conditions are met. According to the Nyquist-Shannon sampling theorem [12], to avoid aliasing, the sampling clock frequency needs to be at least twice the highest clock frequency[3] in the system. In our project, the fastest clock has a frequency of 100 MHz. Hence we need a sampling clock with a frequency of at least 200 MHz, we will use a 300 MHz clock.
2. The Trigger source starts capturing data, based on trigger condition(s). In the case of our simple project, we will trigger on the start button.
3. The Buffer specifies the size of on-chip FPGA memory for data storage. We will use a buffer size of 4K. The sampling frequency, buffer size and number of signals that we monitor determine the amount of time for which data can be stored.

Please refer to the online video on the companion website: http://www.harpgroup. org/muthuswamy/ARouteToChaosUsingFPGAs/ReferenceDesigns/volumeI-Exper

---

[3]Although chaotic systems are mathematically not band-limited, "most" of the chaos is practically band-limited due to the underlying sampling period of the numerical method. Hence a good rule of thumb is to choose the sampling clock frequency to be twice the frequency underlying the numerical method.

Fig. 3.16 Results from the SignalTap session. Notice how the 100 MHz clock has twice the frequency of the 50 MHz clock. Since we are not using a global reset, the signals are not synchronized

imentalObservations/chapter3/SignalTapDemo/. However the result that we should obtain via SignalTap is shown in Fig. 3.16.

3.5.2 Debugging the Chen System Using SignalTap

For understanding how to use SignalTap to debug the Chen system, please refer to the online video and reference design on the companion website: http://www.harpgroup. org/muthuswamy/ARouteToChaosUsingFPGAs/ReferenceDesigns/volumeI-Exper imentalObservations/chapter3/chenSystem/ to understand how to debug the Chen system using SignalTap.

3.6 Hardware Debugging Concepts

Before we conclude this chapter, it would be instructive to discuss some general hardware debugging ideas [1] since the designs that we are going to specify using VHDL are quite complex. We have also discussed debugging via simulation and an in-system logic analyzer in this chapter, so this would be an appropriate place in the book to summarize some engineering debugging ideas. Most of this section has been paraphrased from Altera's online documentation for debugging hardware. If you are using an FPGA from a different manufacturer, you should consult the manufacturer's debugging documentation for additional tips. However this section is quite general and should apply to debugging hardware, irrespective of the FPGA manufacturer.

Debugging of complex logic circuits can be difficult. The task is made easier if one uses an organized approach with the aid of debugging tools. The debugging task involves:

1. Observing that there is a problem
2. Identifying the source of the problem
3. Determining the design changes that have to be made
4. Changing and re-implementing the design
5. Testing the corrected design

3.6.1 Observing a Problem

Often it is easy to see that there is a problem because hardware functionality does
not match the designer's expectations in an obvious way. For example, the graphical
image displayed by the RTL Viewer may indicate that there are missing logic blocks
and/or connections. It is usually a good idea to check at least the top-level of our
design. Consider the RTL view in Fig. 3.17.

Notice that there is no output from the block **hundredth_sec**. The probable reason
is that the outputs of this module are not being used as inputs anywhere else in the
design. Hence the Compiler decided to omit all output signals from this module.
Making this observation via the RTL view would solve the problem.

As another example, suppose the designer assumes erroneously the labeling of
the elements of a VHDL std_logic_vector. Consider for instance the labeling of the
segments for a seven segment display. Compiling, synthesizing and programming
the FPGA would result in a circuit that seems to respond properly to pressing of
input keys, but generates a strange looking output on the seven segment displays.
Observing this behavior, the designer may suspect that there is something wrong with
the decoder itself. A simple test is to use ModelSim or the SignalTap logic analyzer
to ensure that the outputs of the decoder are in the correct order.

A complex design may be difficult to debug. The implementation may appear to
contain all necessary components, it may appear to function properly, but the results
it produces do not exhibit the expected behavior. In such cases, the first task is to
correctly identify the source of the problem.

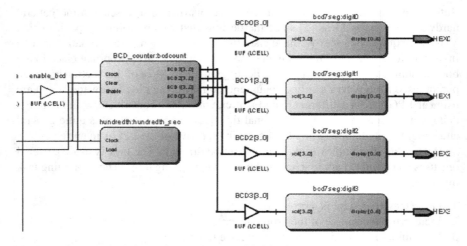

Fig. 3.17 The erroneous circuit displayed by the RTL Viewer

3.6.2 Identifying the Problem

Designer's intuition (which improves greatly with experience) may suggest some
tests that could be tried. Otherwise, it is necessary to adopt an organized procedure.
A golden rule is to first test small portions of the circuit, which should be easy to do
if the circuit is designed in a modular fashion. This is referred to as the divide-and-
conquer approach.

In this book, you should have noticed that we do emphasize the modular approach.
Specifically, the top-level of our design in the RTL Viewer should show only blocks
and tri-state buffers. For instance, consider the DE2 Chaos Engine that implements a
variety of chaotic systems, found online: http://www.harpgroup.org/muthuswamy/
ARouteToChaosUsingFPGAs/ReferenceDesigns/volumeI-ExperimentalObservatio
ns/DE2ChaosEngine.zip.

The Compilation Report for the design in Fig. 3.18 is shown in Fig. 3.19.

Notice how the modular top-level has abstracted away a quite complicated design.
Moreover, the modular approach allows a designer to compile, simulate and test each
module on its own. We have also been emphasizing this approach when we simulated
the discrete Euler method in MATLAB before FPGA specification.

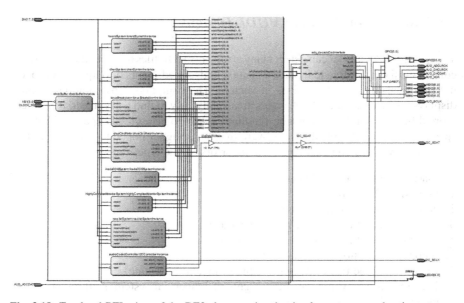

Fig. 3.18 Top-level RTL view of the DE2 chaos engine that implements seven chaotic systems:
Lorenz, Rössler, Highly Complex Attractor, Chen, Ikeda Delay Differential Equation, Chua oscillator and Torus Breakdown

Flow Summary	
Flow Status	In progress - Sun Mar 23 23:05:16 2014
Quartus II 32-bit Version	12.0 Build 178 05/31/2012 SJ Full Version
Revision Name	DE2ChaosEngine
Top-level Entity Name	DE2ChaosEngine
Family	Cyclone IV E
Device	EP4CE115F29C7
Timing Models	Final
• Total logic elements	69,370 / 114,480 (61 %)
Total combinational functions	59,773 / 114,480 (52 %)
Dedicated logic registers	38,575 / 114,480 (34 %)
Total registers	38575
Total pins	104 / 529 (20 %)
Total virtual pins	0
Total memory bits	85,698 / 3,981,312 (2 %)
Embedded Multiplier 9-bit elements	531 / 532 (100 %)
Total PLLs	1 / 4 (25 %)

Fig. 3.19 Compilation report for the DE2 chaos engine. It took 45 min to assemble a programmable file on Dr. Muthuswamy's Windows 7 emulator under Parallels Desktop

3.6.3 Sources of Errors in VHDL Designs

The Quartus II Compiler can detect many errors in VHDL files that specify a given circuit. Typical errors include incorrect syntax, undeclared inputs or outputs, improper use of signals and incorrect sizes of vectors. The compiler stops compilation and displays an error message. Such errors are usually easy to find and correct. It is much more difficult to find errors in a design that appears to be correctly specified but the specification does not result in hardware that the designer hoped to achieve. In this subsection, we will consider some typical errors of this type:

1. Inadvertent creation of latches
2. Omission of signals
3. Not assigning a value to a wire
4. Assigning a value to a wire more than once
5. Incorrect specification of **port map** signals
6. Incorrect definition of signal vector
7. Incorrectly specified FSM (e.g. wrong or invalid next state)
8. Incorrect timing where the output signal of a given subsystem is off by one clock cycle
9. Careless use of clocks

Inadvertent latches are created by the Compiler if the designer fails to specify the action needed for all cases in constructs. Latches can also be inferred by the Compiler if the designer forgets to register input(s) on clock edges.

If the designer fails to use some signals in a VHDL file, the Compiler will ignore these signals completely and may even omit circuitry associated with these signals. Failure to include the **begin** and **end** delimiters in a multi-statement **process** block will cause only one statement to be considered valid. Careful use of blocking and nonblocking assignments is essential. It is dangerous, and not advisable, to use both types of assignments in the same **process** block. To describe a combinational circuit in a **process** construct, it is best to use blocking assignments. For sequential circuits, it is best to use nonblocking assignments.

Errors in the specification of an FSM may lead to a variety of undesirable consequences. They can cause wrong functional behavior by reaching wrong states, as well as wrong timing behavior by producing incorrect output signals. A common error results in an output signal that is off by one clock cycle.

It is particularly important to use clocks carefully. We already discussed in Sect. 2.3.2.4 the implications of having a single synchronous clock. If we have to utilize different clock frequencies for interfacing to external components, it is best to use PLL(s).

Inadequate understanding of the board can lead to design errors. Typical examples include:

1. Wrong pin assignment
2. Wrong interpretation of the polarity of pushbutton keys and toggle switches
3. Timing issues when accessing various chips on the board, such as SDRAM memory

If pins are not assigned correctly, the design will not exhibit desired behavior. The Quartus II compiler causes all unused pins to be driven to ground by default. The easiest way of ensuring that the pins are correctly assigned for the board is to import the (usual) manufacturer provided pin assignment file. If the design involves access to external peripherals (particularly memory like SDRAM), it is necessary to adhere to strict timing requirements by utilizing the peripheral's data sheet.

3.6.4 Design Procedure

It is prudent to follow a systematic design procedure that tends to minimize the number of design errors and simplifies the debugging tasks. Here are final suggestions that are likely to help:

1. Design the system in a modular, hierarchical manner.
2. Use well-understood and commonly-used constructs to define circuits.
3. Test each module, by simulating it, before it is incorporated into a larger system.
4. Define and test portions of the final design by connecting two or more modules.
5. If possible, simulate the full design.

It is vital that we write VHDL in a style that allows one to easily visualize the hardware specified by the code. It is also useful to make the resulting VHDL easily understandable for others.

3.7 Another Example—A Highly Complex Attractor System

In Sect. 3.2.1, we realized the classic Lorenz system. In this section, we will realize a system that will show the robustness of using DSP builder. Consider the Eqs. (3.10), (3.11) and (3.12) [10].

$$\dot{x} = y - x \tag{3.10}$$

$$\dot{y} = -\tanh(x)z \tag{3.11}$$

$$\dot{z} = -R + xy + |y| \tag{3.12}$$

$R = 60$ in Eq. (3.12) and initial conditions are $(1, 1, 1)$. The robustness of DSP builder will be evident when we are able to realize the hyperbolic tangent function in Eq. (3.11). What makes this system interesting is the fact that as of 2013, this system has the largest Lyapunov dimension of 2.3 (for a dissipative chaotic system) [10]. In this section, we will first simulate and then realize this system on the FPGA.

Figures 3.20 and 3.21 show a portion of the Simulink block diagram and $x(t)$ from the simulation result respectively.

Notice that due to the large Lyapunov dimension, our simulation parameters need to closely match our implementation sampling frequency and dt. The step size used in simulation implies large amounts of memory (we generate 40000001 x, y, z

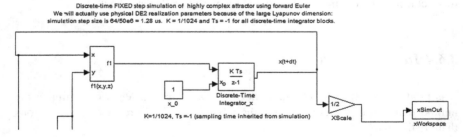

Fig. 3.20 A screenshot of the Simulink discrete Euler realization of the highly complex attractor system. You can download Simulink design from the companion website: http://www.harpgroup.org/muthuswamy/ARouteToChaosUsingFPGAs/ReferenceDesigns/volumeI-ExperimentalObservations/chapter3/highlyComplexAttractor/

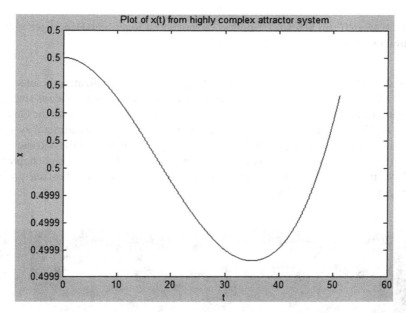

Fig. 3.21 $x(t)$ from the highly complex attractor system. Notice how the large Lyapunov dimension affects our simulation in the sense we need a large simulation time step to save memory

values). One solution to overcome the memory issue is to run a hardware functional simulation.[4]

Example 3.1 Implement $\tanh(x)$ in DSP Builder Advanced blockset

Solution: Upon examining the ModelPrim blocks, we will notice that we do not have a hyperbolic tangent function. Nevertheless, using first principles, we get:

$$\tanh(x) = \frac{\sinh(x)}{\cosh(x)} \tag{3.13}$$

$$= \frac{0.5(e^x - e^{-x})}{0.5(e^x + e^{-x})} \tag{3.14}$$

$$= \frac{e^{2x} - 1}{e^{2x} + 1} \tag{3.15}$$

---

[4]We could also reduce simulation step size and K in the Simulink simulation but you will notice our results will not match with the physical realization (when compared with the Lorenz system) because of the large Lyapunov dimension. Turns out our choices for sampling frequency and dt are sufficient to simulate the system accurately. But before we look at the physical realization, we need to address the issue of specifying the hyperbolic tangent function in DSP builder.

In the ModelPrim blocks, we do have exponential functions. Hence, we will implement Eq. (3.15) in DSP Builder.

Figures 3.22, 3.23, 3.24, 3.25 and 3.26 show physical realization results. The VHDL specification is in listing D.5. We have however not recorded videos since the concepts are analogous to the Lorenz system. You can however download the Quartus project that incorporates the Lorenz system and the highly complex attractor system[5] from the companion website: http://www.harpgroup.org/muthuswamy/ARouteToCh aosUsingFPGAs/ReferenceDesigns/volumeI-ExperimentalObservations/chapter3/. We have not synthesized the Quartus project since the fully synthesized project size is approximately 90 MB.

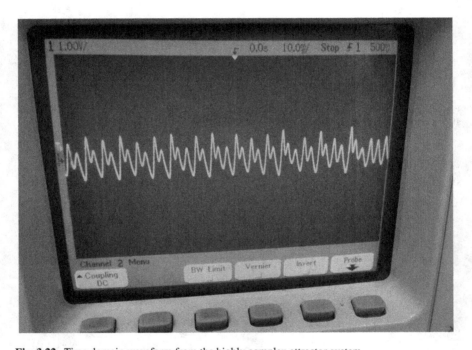

Fig. 3.22 Time domain waveform from the highly complex attractor system

---

[5]There are other chaotic systems in the zipped Quartus project that we will utilize later in the book.

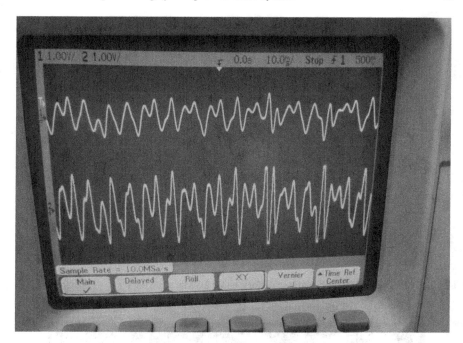

Fig. 3.23 The two other state variables from the highly complex attractor system plotted in the time-domain

Fig. 3.24 $y(t)$ versus $x(t)$, 0.5 V/div scale for y and 0.2 V/div scale for x. We used the analog Tektronix 2205 scope to capture XY waveforms because, due to the large Lyapunov dimension, the features of the attractor in phase space were better resolved in an analog scope

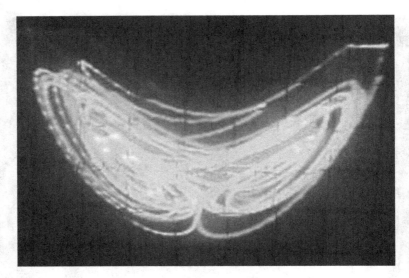

Fig. 3.25 $z(t)$ versus $x(t)$, 0.5 V/div scale for z and 0.2 V/div scale for x

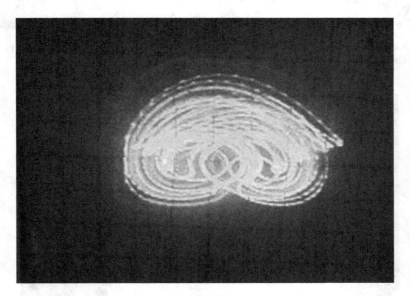

Fig. 3.26 $z(t)$ versus $y(t)$, 0.5 V/div scale for both axes

3.8 Conclusions

Below is a summary of the main concepts and steps used in this chapter:

1. We have made a design choice of specifying synchronous Euler's method in VHDL instead of in DSP Builder, please refer to listing D.4.
2. We should first run a discrete-time simulation of our chaotic differential equation to identify ideal sampling frequencies and to get an idea of the maximum (minimum) state variable values so we can appropriately scale the output to match the DAC range of our audio codec.
3. We need to add both the nonlinear subsystem and nonlinear synthesizable subsystem from our DSP builder design to our Quartus project. We also need to add the following DSP builder libraries (from the appropriate Quartus installation directory):

```
altera/12.0/quartus/dspba/Libraries/vhdl/fpc/math_package.vhd
altera/12.0/quartus/dspba/Libraries/vhdl/fpc/math_implementation.vhd
altera/12.0/quartus/dspba/Libraries/vhdl/fpc/hcc_package.vhd
altera/12.0/quartus/dspba/Libraries/vhdl/fpc/hcc_implementation.vhd
altera/12.0/quartus/dspba/Libraries/vhdl/fpc/fpc_library_package.vhd
altera/12.0/quartus/dspba/Libraries/vhdl/fpc/fpc_library.vhd
```

4. We need to add fixed point wrapper functions [5] around our DSP builder synthesized subsystem so that we can use the audio codec controller, please refer to listing D.4. Note that the number of bits for the fixed point integer and decimal part will vary with the chaotic system.
5. A ModelSim simulation of our chaotic system is used to confirm the functionality of our digital design. Since the discrete Euler's method in Simulink cannot take into account our entire digital system (including the i^2c protocol for the audio codec), a ModelSim simulation is a must.
6. A test bench should be utilized to test the different sub-components of our design. Test benches are not synthesizable, so any VHDL constructs can be used.
7. In-system debugging provides an alternative to timing intensive simulation. Moreover, we are debugging the design as it is executing on the FPGA.
8. Debugging is both an art and a science. But remember this quote from Einsten while debugging: "Insanity: doing the same thing over and over again but expecting different results".

We now have the necessary FPGA knowledge to study a very important concept that is central to the formation of a chaotic attractor-bifurcations. This is the subject of Chap. 4.

Problems

3.1 Experiment with the sampling frequency and fixed point representation for the highly complex attractor system.

3.2 Reconsider the Chua system (Eqs. 1.82, 1.83 and 1.84), repeated below for convenience.

$$\dot{x} = \alpha[y - x - m_1 x - \frac{1}{2}(m_0 - m_1)(|x + 1| - |x - 1|)] \qquad (3.16)$$

$$\dot{y} = x - y + z \qquad (3.17)$$

$$\dot{z} = -\beta y \qquad (3.18)$$

$m_0, m_1, \alpha, \beta \in \mathbb{R}$ are parameters of the system. Use $m_0 = \frac{-8}{7}, m_1 = \frac{-5}{7}$, $\alpha = 15.6, \beta = 25.58$ to perform a discrete simulation and then realize the system on the FPGA.

3.3 Simulate and implement Sprott's jerky chaotic system shown in Eqs. (3.19), (3.20) and (3.21).

$$\dot{x} = -2y \qquad (3.19)$$

$$\dot{y} = x + z^2 \qquad (3.20)$$

$$\dot{z} = 1 + y - 2z \qquad (3.21)$$

3.4 Simulate and implement the Rössler system in Eqs. (3.22), (3.23) and (3.24).

$$\dot{x} = -y - z \qquad (3.22)$$

$$\dot{y} = x + \alpha y \qquad (3.23)$$

$$\dot{z} = \beta + z(x - \gamma) \qquad (3.24)$$

$\alpha, \beta, \gamma \in \mathbb{R}$ are parameters of the system. Use $\alpha = 0.1, \beta = 0.1, \gamma = 14$.

3.5 Simulate and implement the hyperchaotic Lü system [3] in Eqs. (3.25), (3.26), (3.27) and (3.28).

$$\dot{x} = a(y - x) + u \qquad (3.25)$$

$$\dot{y} = -xz + cy \qquad (3.26)$$

$$\dot{z} = xy - bz \qquad (3.27)$$

$$\dot{u} = xz + du \qquad (3.28)$$

Parameter values are: $a = 36, b = 3, c = 20, d = 1.3$. Initial conditions are $(1, 2, 3, 4)$. Note that since we have a four-dimensional system and for the parameters chosen we have a value of hyperchaos. When we realize this system on the FPGA, we have to make sure that we output all four state variables via the DAC on the audio codec.

3.6 We could make a refinement on the simple forward-Euler numerical method for solving chaotic differential equations by considering the fourth-order RK method shown in Eqs. (3.29)–(3.33) [11].

$$\mathbf{k}_1 = \mathbf{F}(\mathbf{x}_t)\delta t \tag{3.29}$$

$$\mathbf{k}_2 = \mathbf{F}\left(\mathbf{x}_t + \frac{\mathbf{k}_1}{2}\right)\delta t \tag{3.30}$$

$$\mathbf{k}_3 = \mathbf{F}\left(\mathbf{x}_t + \frac{\mathbf{k}_2}{2}\right)\delta t \tag{3.31}$$

$$\mathbf{k}_4 = \mathbf{F}\left(\mathbf{x}_t + \mathbf{k}_3\right)\delta t \tag{3.32}$$

$$\mathbf{x}_{t+\delta t} = \mathbf{x}_t + \frac{\mathbf{k}_1}{6} + \frac{\mathbf{k}_2}{3} + \frac{\mathbf{k}_3}{3} + \frac{\mathbf{k}_4}{6} \tag{3.33}$$

Implement Eqs. (3.29)–(3.33) on the FPGA. A maximum step size of $\delta t = 0.1$ is more than adequate for most cases because the natural period of oscillation is typically less than 1 rad per second when the parameters are of order unity, and thus there is the order of $\frac{2\pi}{\delta t} \approx 63$ iterations per cycle [11]. Using this idea, determine an appropriate step size (Δt) and sampling count for the FPGA realization. Test your algorithm by implementing the circulant chaotic system [11] on the FPGA.

$$\dot{x} = -ax + by - y^3 \tag{3.34}$$

$$\dot{y} = -ay + bz - z^3 \tag{3.35}$$

$$\dot{z} = -az + bx - x^3 \tag{3.36}$$

Parameters are: $a = 1, b = 4$. Initial conditions are $(0.4, 0, 0)$. Note that the online reference design for this chapter has an Euler's method realization of the circulant chaotic system. Compare your RK realization to the Euler realization.

3.7 Perform a functional simulation of the simple combinational logic design from Sect. 2.3.2.1 (listing 2.1). Note that although a functional simulation is "overkill" for this design, this problem should help you understand the nuances of ModelSim by using a very simple example for simulation.

3.8 Perform a functional simulation of the generic n-bit ripple carry adder from Sect. 2.3.2.3.

3.9 Perform a functional simulation of the i²c design from Sect. 2.3.3.

3.10 Design and verify the SignalTap waveforms for the PLL based design in Sect. 3.5.1. Also add a global asynchronous reset.

Lab 3 : ModelSim Simulation, In-System Debugging and Physical Realization of the Muthuswamy-Chua System

Objective: Simulate and physically realize the Muthuswamy-Chua [6, 8, 13] system.

Theory: The Muthuswamy-Chua system models a linear inductor, linear capacitor and memristor in series (or parallel). In this book, we will use the original series version whose system equations are Eqs. (3.37)–(3.39) [8].

$$\dot{v}_C = \frac{i_L}{C} \tag{3.37}$$

$$\dot{i}_L = \frac{-1}{L}\left(v_C + R(z)i_L\right) \tag{3.38}$$

$$\dot{z} = f(z, i_L) \tag{3.39}$$

Lab Exercise:

Simulate, verify using SignalTap and hence implement Eqs. (3.37)–(3.39) for the following parameters and functions: $C = 1, L = 3, R(z) = \beta(z^2 - 1), f(z, i_L) = -i_L - \alpha z + z i_L, \beta = 1.5, \alpha = 0.6$. Initial conditions are: $(0.1, 0, 0.1)$. For the simulation, please check the output from the audio codec controller as well. This will help you determine if the left-channel and right-channel DAC registers use the entire range of values reserved for 16-bit 2's complement.

References

1. Altera Corporation (2014) Debugging of VHDL Hardware Designs on Altera's DE-Series Boards.   ftp://ftp.altera.com/up/pub/Altera_Material/13.0/Tutorials/VHDL/Debugging_ Hardware.pdf 23 Mar 2014
2. Altera Corporation (2014) Quartus II 10.0 handbook—Design Debugging Using the SignalTap II Logic Analyzer. http://www.altera.com/literature/hb/qts/qts_qii53009.pdf 23 Mar 2014
3. Chen A et al (2006) Generating hyperchaotic Lü attractor via state feedback control. Phys A 364:103–110
4. Chu PP (2011) Embedded SOPC design with NIOS II processor and VHDL examples. Wiley, Hoboken
5. Cornell University (2013) Digital Differential Analyzer. In: ECE5760 Homepage. http://people. ece.cornell.edu/land/courses/ece5760/DDA/index.htm Accessed 13 Oct 2013
6. Llibre J, Valls C (2012) On the integrability of a Muthuswamy-Chua system. J Nonlinear Math Phys 19(4):1250029–1250041
7. Lu J et al (2002) Local bifurcations of the Chen system. Int J Bifurc Chaos 12(10):2257–2270
8. Muthuswamy B, Chua LO (2010) Simplest chaotic circuit. Int J Bifurc Chaos 10(5):1567–1580
9. Numerical Simulation of Chaotic ODEs (2014) In: Chaos from Euler Solution of ODEs. http:// sprott.physics.wisc.edu/chaos/eulermap.htm 24 Aug 2014
10. San-Um W, Srisuchinwong B (2012) Highly complex chaotic system with piecewise linear nonlinearity and compound structures. J Comput 7(4):1041–1047
11. Sprott JC (2010) Elegant chaos. World Scientific, New Jersey
12. Varaiya PP, Lee EA (2002) Structure and interpretation of signals and systems. Addison-Wesley, Boston
13. Zhang Y, Zhang X (2013) Dynamics of the Muthuswamy-Chua system. Int J Bifurc Chaos 23(8):1350136–1350143

Chapter 4
Bifurcations

FPGA realization of the torus breakdown system [6]

Abstract This chapter will explore a variety of routes that lead to chaos in dynamical systems, through simulation and FPGA experiments. The goal of this chapter is simply for the reader to understand that a system is chaotic for a certain range of parameters and there are interesting mechanisms that lead to the chaotic behavior.

4.1 The Concept of Bifurcations

Simply put, bifurcations are sudden qualitative changes in system dynamics at critical parameter values [1]. For example, consider the Rössler system in Problem 2 from Chap. 1.

$$\dot{x} = -y - z \tag{4.1}$$

$$\dot{y} = x + \alpha y \tag{4.2}$$

$$\dot{z} = \beta + z(x - \gamma) \tag{4.3}$$

We will first specify the system from Eqs. (4.1)–(4.3) in MATLAB. In listing E.1, we have utilized parameterized functions so that we can pass in ODE simulation settings and parameters. On the MATLAB command line, we can type listing E.2 to obtain a phase plot, with parameter values $\alpha = 0.1$, $\beta = 0.1$ and $\gamma = 4$. In listing E.2, we only plot a subset of our state variables to account for transient behavior. Figure 4.1 shows the result.

© Springer International Publishing Switzerland 2015

B. Muthuswamy and S. Banerjee, *A Route to Chaos Using FPGAs*, Emergence, Complexity and Computation 16, DOI 10.1007/978-3-319-18105-9_4

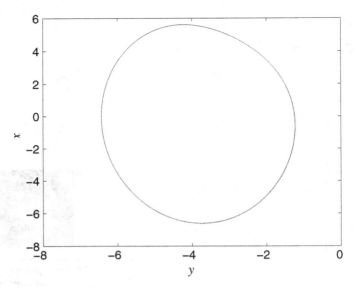

Fig. 4.1 y versus x phase plot for the Rössler system from listings E.1 and E.2

We will now investigate the behavior of our system as we change one of the parameters while keeping the other parameters fixed. We will use phase space to visualize the change in system behavior, leading to chaos. This "birth" of chaos is technically called as "route to chaos".

Note that the question regarding the most common route to chaos is, in any but a very select set of specific examples, still an open and poorly defined question [2]. Even analytically piecing together the types of bifurcations that can exist en route to chaos is usually difficult.

4.2 Routes to Chaos

There are a variety of routes to chaos, in this section we will discuss some of the more "common" types and investigate bifurcation behaviour using MATLAB. FPGA investigation of bifurcations is covered in Sect. 4.3.

4.2.1 Period-Doubling Route to Chaos

In this route to chaos, an equilibrium point looses stability and a stable limit cycle emerges through an Andronov-Hopf bifurcation [3]. As we continue changing the value of a parameter, a stable limit cycle at approximately twice the period emerges, which we will refer to as a period-2 limit cycle. As the parameter value is further changed, the period-2 limit cycle in turn loses stability and a stable period-4 limit

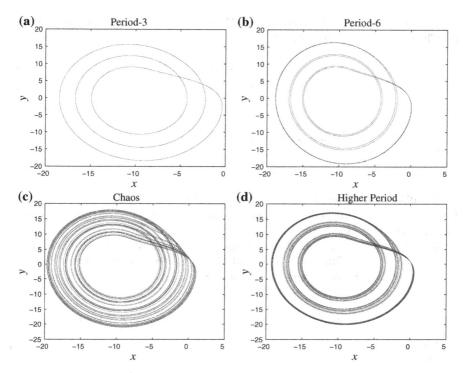

Fig. 4.2 Period-doubling route to chaos in the Rössler system. $\alpha = 0.1$, $\beta = 0.1$ for all the plots above. Bifurcation parameter value of γ (starting clockwise from *top-left*): Period-3—$\gamma = 12$; Period-6—$\gamma = 12.6$; Higher Period—$\gamma = 13.3$; Chaos—$\gamma = 14$

cycle appears. This bifurcation occurs many times at ever decreasing intervals of the parameter range which converges at a geometric rate to a limit when chaos is observed.

Let us use the Rössler system to illustrate period-doubling. Listing E.3 gives MATLAB code, results are shown in Fig. 4.2.

Notice that it becomes quite difficult to visually spot the period (Fig. 4.2d) as the system tends towards chaotic behaviour.

4.2.2 Period-Adding Route to Chaos

In this route to chaos, we will have windows of consecutive periods separated by regions of chaos [3]. In other words as the parameter is varied, we obtain a stable period-n orbit, $n = 1, 2, \ldots$ followed by a region of chaos, then a stable period-$(n+1)$ orbit, followed by chaos and then a period-$(n + 2)$ orbit and so on.

Consider Chua's oscillator equations [4] in Eqs. (4.4)–(4.6), a generalization of the Chua system from Problem 1.7.

$$\dot{x} = \alpha(y - g(x)) \tag{4.4}$$

$$\dot{y} = x - y + z \tag{4.5}$$

$$\dot{z} = -\beta y - \gamma z \tag{4.6}$$

The nonlinear function g is given in Eq. (4.7).

$$g(x) = ax^3 + cx \tag{4.7}$$

The difference between Chua's oscillator and Chua's circuit is the γz term in Eq. (4.6). In the physical circuit realization, this term is obtained by using a resistor in series with the inductor [4].

In order to observe period-adding route to chaos, we will fix parameters $\alpha = 3.708, \gamma = 0.076, a = 1, c = -0.276$. β will be varied. Listings E.4 and E.5 show MATLAB code, results are in Fig. 4.3.

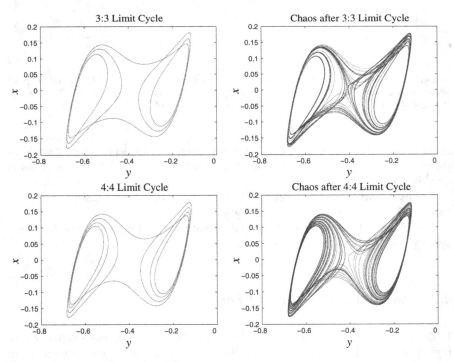

Fig. 4.3 Period-adding route to chaos, obtained using Chua's oscillator. For 3:3 limit cycle $\beta = 3.499$, for chaos after 3:3 limit cycle $\beta = 3.708$, for 4:4 limit cycle $\beta = 3.574$ and for chaos after 4:4 limit cycle $\beta = 3.6$

4.2.3 Quasi-Periodic Route to Chaos

In this route to chaos, we will have a toroidal attractor bifurcating into a chaotic attractor [3]. The toroidal attractor is initially formed due to two incommensurate frequencies.[1] Consider the system of equations in Eqs. (4.8)–(4.10) [5].

$$\dot{x} = -\alpha f(y - x) \tag{4.8}$$

$$\dot{y} = -f(y - x) - z \tag{4.9}$$

$$\dot{z} = \beta y \tag{4.10}$$

The nonlinear function f is given in Eq. (4.11).

$$f = -ax + \frac{1}{2}(a + b)\,(|x + 1| - |x - 1|) \tag{4.11}$$

The bifurcation sequence is shown in Fig. 4.4, the relevant code is in listings E.6 and E.7. Here, we have a two-torus, namely a quasi-periodic solution and as we increase α further we observe that the two-torus and the periodic attractor (phase-locking) alternatively appear and disappear many times [5]. As we continue to increase α, the two-torus will fail to appear and we can obtain chaos through the period-doubling, period-adding (discussed earlier) or torus-breakdown. This section discussed torus-breakdown, the other two scenarios are left as exercises.

4.2.4 Intermittency Route to Chaos

Intermittency is the phenomenon where the signal is virtually periodic except for some irregular (unpredictable) bursts [3]. In other words, we have intermittently periodic behaviour and irregular aperiodic behaviour.

In this section, we will start with the physical Chua's oscillator [3], to illustrate the idea of dimensionless scaling.[2] We will then choose a set of parameter values such that the intermittency route to chaos is observed.

> *Example 4.1* Consider the circuit equations of Chua's oscillator shown in Eqs. (4.12)–(4.14). Scale the circuit equations into a dimensionless form suitable for implementation on an FPGA [6].

---

[1] Incommensurate frequencies ω_1 and ω_2 imply that the ratio $\frac{\omega_1}{\omega_2} \in \mathbb{R}\backslash\mathbb{Q}$.

[2] Although dimensionless scaling could have been covered in Chap. 1, we delayed introducing this concept so the reader can appreciate the idea better, once they have been well-exposed to chaos.

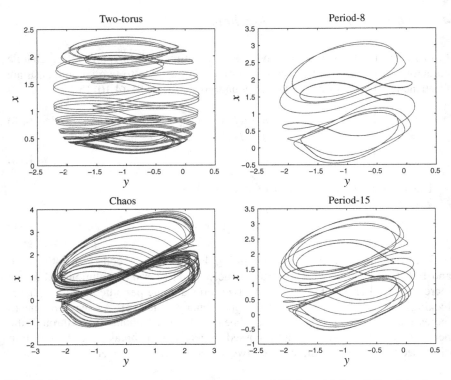

Fig. 4.4 Torus-breakdown route to chaos. $\beta = 1, a = 0.07, b = 0.1$ for all the plots above. Bifurcation parameter value of α (starting clockwise from *top-left*): Two-torus—$\alpha = 2.0$; Period-8—$\alpha = 8.0$; Period-15—$\alpha = 8.8$; Chaos—$\alpha = 15.0$

$$\dot{v}_1 = \frac{1}{C_1}\left(\frac{(v_2 - v_1)}{R} - f(v_1)\right) \tag{4.12}$$

$$\dot{v}_2 = \frac{1}{C_2}\left(\frac{(v_1 - v_2)}{R} + i_3\right) \tag{4.13}$$

$$\frac{di_3}{dt} = -\frac{1}{L}(v_2 + R_0 i_3) \tag{4.14}$$

The piecewise-linear function f in Eq. (4.12) is given in Eq. (4.15):

$$f(v_1) = G_b v_1 + \frac{1}{2}(G_a - G_b)(|v_1 + E| - |v_1 - E|) \tag{4.15}$$

Solution: Notice we know have a piecewise linear function for the nonlinearity, as opposed to Eq. (4.7) in Sect. 4.2.2. In order to perform dimensionless scaling, we need to cast voltage, current and time into dimensionless form. To do so consider the following definitions:

$$x \overset{\Delta}{=} \frac{v_1}{E}, \qquad y \overset{\Delta}{=} \frac{v_2}{E}, \qquad z \overset{\Delta}{=} i_3 \frac{R}{E}, \qquad \tau \overset{\Delta}{=} \frac{t}{|RC_2|} \tag{4.16}$$

$$\alpha \overset{\Delta}{=} \frac{C_2}{C_1}, \qquad \beta \overset{\Delta}{=} \frac{R^2 C_2}{L}, \qquad \gamma \overset{\Delta}{=} \frac{RR_0 C_2}{L} \tag{4.17}$$

$$a \overset{\Delta}{=} RG_a, \qquad b \overset{\Delta}{=} RG_b, \qquad k = 1, \text{ if } RC_2 > 0, \qquad k = -1, \text{ if } RC_2 < 0 \tag{4.18}$$

Utilizing the definitions in Eqs. (4.16)–(4.18), we get Eqs. (4.19)–(4.22).

$$\frac{dx}{d\tau} = k\alpha \, (y - x - f(x)) \tag{4.19}$$

$$\frac{dy}{d\tau} = k \, (x - y + z) \tag{4.20}$$

$$\frac{dz}{d\tau} = k \, (-\beta y - \gamma z) \tag{4.21}$$

$$f(x) = bx + \frac{1}{2}(a - b) \, (|x + 1| - |x - 1|) \tag{4.22}$$

Parameter values for the intermittency route to chaos are $\alpha = -75.018755$, $a = -0.98$, $b = -2.4$, $k = 1$. In this case, we will have two bifurcation parameters: β and γ. However the intermittency route to chaos is nevertheless a co-dimension one bifurcation in the sense that the corresponding route in the parameter space is a 1-D curve [7].

Simulation code is shown in listings E.8 and E.9. Simulation result is shown in Fig. 4.5.

4.2.5 Chaotic Transients and Crisis

Transient chaos is the mechanism by which a trajectory typically behaves chaotically for a finite amount of time before settling into a final (usually nonchaotic state) [8]. The dynamical origin of transient chaos is known: it is due to nonattracting chaotic saddles in phase space [8]. What is interesting about transient chaos is that we have chaotic saddles, unlike say the Lorenz system where we have chaotic attractors. A chaotic attractor is a bounded set that exhibits a fractal structure only in the stable

Fig. 4.5 The intermittency route to chaos. Bifurcation parameter value for β, γ (starting from *top*): Periodic— $\beta = 44.803$, $\gamma = -4.480$; First chaotic intermittency— $\beta = 43.994721$, $\gamma = -4.3994721$; Second chaotic intermittency— $\beta = 31.746052$, $\gamma = -3.1746032$; Third chaotic intermittency— $\beta = 31.25$, $\gamma = -3.125$

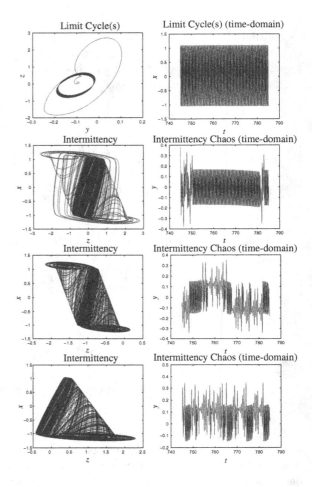

direction whereas a chaotic saddle is a bounded set that exhibits a fractal structure in both stable and unstable directions.[3] Due to the fractal structure in the unstable direction, an infinite number of gaps of all sizes exists along the unstable manifold of the chaotic saddle. An initial condition is typically attracted toward the chaotic saddle along the stable direction, stays in its vicinity for a finite amount of time, and then leaves the chaotic saddle through one of the gaps in the unstable direction. It is known that chaotic saddles and transient chaos are responsible for important physical phenomena such as chaotic scattering and particle transport in open hydrodynamical flows [8]. We will now consider an example of chaotic transients in a physical model, that will lead to species extinction!

---

[3]Detailed theoretical methods to understand chaotic systems as pertaining to FPGA realizations will be covered in Volume II.

Extinction of species has been one of the biggest mysteries in nature [8]. A common belief about local extinction is that it is typically caused by external environmental factors such as sudden changes in climate. For a species of very small population size, small random changes in population (known as "demographic stochasticity") can also lead to its extinction. Clearly, the question of how species extinction occurs is extremely complex, as each species typically lives in an environment that involves interaction with many other species (e.g., through competition for common food sources, predator-prey interactions, etc.) as well as physical factors such as weather and disturbances. From a mathematical point of view, a dynamical model for the population size of a species is complex, involving both spatial and temporal variations. Thus such a system in general should be modeled by nonlinear partial differential equations. An obvious difficulty associated with this approach is that the analysis and numerical solution of such nonlinear partial differential equations present an extremely challenging problem in mathematics.

Nonetheless, in certain situations, the problem of species extinction may become simpler. Specifically, in this section, we will use the much simpler three dimensional nonlinear ODEs suggested by McCann and Yodzis [8]: a resource species, a prey (consumer) and a predator. The population densities of these three species denoted by R, C and P for resource, consumer and predator, respectively, are governed by Eqs. (4.23)–(4.25).

$$\frac{dR}{dt} = R\left(1 - \frac{R}{K}\right) - \frac{x_C y_C C R}{R + R_0} \tag{4.23}$$

$$\frac{dC}{dt} = x_C C\left(\frac{y_C R}{R + R_0} - 1\right) - \frac{x_P y_P P C}{C + C_0} \tag{4.24}$$

$$\frac{dP}{dt} = x_P P\left(-1 + \frac{y_P C}{C + C_0}\right) \tag{4.25}$$

K is the resource carrying capacity and x_C, y_C, x_P, y_P, R_0 and C_0 are parameters that are positive. The model carries the following biological assumptions [8]:

1. The life histories of each species involve continuous growth and overlapping generations, with no age structure (this permits the use of differential equations)
2. The resource population (R) grows logistically
3. Each consumer species (immediate consumer C, top consumer P) without food dies of exponentially
4. Each consumer's feeding rate (example, $\frac{x_C y_C R}{R + R_0}$) saturates at high food levels

Realistic values for parameters can be derived from bioenergetics. Following [8], we fix $x_C = 0.4$, $y_C = 2.009$, $x_P = 0.08$, $y_P = 2.876$, $R_0 = 0.16129$, $C_0 = 0.5$. The resource carrying capacity, K, can be different in different environments and hence is our bifurcation parameter.

Figure 4.6 show a chaotic attractor and a limit cycle for $K < K_c \approx 0.99976$. There is a period-doubling cascade to chaos and a crisis at $K = K_c$. Note that none of the populations will become extinct for $K < K_c$ because the chaotic attractor is

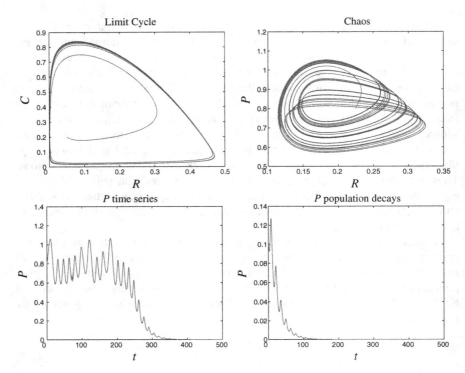

Fig. 4.6 Chaotic Transients and Crisis. Starting from *top*, clockwise: The limit cycle co-exists with the chaotic attractor. Notice that before the onset of crisis, *P* time series decays. The limit cycle and chaotic attractor were obtained by using different sets of initial conditions (for limit cycle: (0.1, 0.2, 0.1), for chaos: (0.55, 0.35, 0.8)), but the same parameter values. After the onset of crisis, even though *P* oscillates chaotically initially, the predator population eventually decays to 0

located in a phase-space region away from the origin. Initial conditions do tend to a trajectory along the second co-existing attractor, the limit cycle. This correspond to the situation where the top predator population becomes extinct. As the carrying capacity increases beyond the critical value K_c, the predator population becomes extinct for almost all initial conditions. This can be understood from dynamical systems theory because at $K = K_c$ a crisis occurs since the tip of the chaotic attractor touches the basin boundary [8], after which there is transient chaos shown in Fig. 4.6. It can be seen that $P(t)$ remains finite initially but decreases rapidly to zero. Thus we see that species extinction can indeed occur as a result of transient chaos.

Simulation code is in listings E.10 and E.11.

4.3 Bifurcation Experiments with an FPGA

The next step is to physically study bifurcation mechanisms in chaotic systems using FPGAs. Most development boards (including the DE2) have hardware debounced push-buttons. These pushbuttons can be used to increment or decrement parameters. As we discussed in Chap. 2, we have four push-buttons on the DE2 board. Since we have used KEY(0) as global reset, we can utilize the rest of the keys for bifurcation experiments. There are two steps involved:

1. Implement a pulse generator that accounts for latency (propagation delay)
2. Implement parameter change(s) (increment, decrement, etc.) using the appropriate floating point modules (adder, subtractor, etc.).

We will illustrate both steps above using the bifurcation scenarios from Sect. 4.2. We will also illustrate the concept of hierarchical design by using subsystems in Simulink for implementing the various nonlinearities.

Example 4.2 Discuss why we need a pulse generator and utilize the single pulse generator from listing C.4 as a reference design, but implement the pulse generator as a Moore FSM.

Solution: The sampling clock period is 20 ns. But pressing and releasing any push button has a minimum human reaction time of 1 ms. Therefore in order to ensure that a single key press is not misinterpreted as multiple key presses, we need a pulse generator. We will also need to account for latency associated with floating point computation. Implementation of a pulse generator Moore FSM that incorporates latency is shown in listing E.12.

   We use a pulse that is eight clock cycles long because the latency associated with the floating point addition or subtraction is seven clock cycles. We wait one more clock cycle to make sure data has been correctly updated on the rising edge of the clock for the floating point module.

Before we move on to the next example of realizing parameter increment, it is a good idea to check our FSM functionality using ModelSim. Listing E.13 shows the test bench, listing E.14 shows the ModelSim script file and Fig. 4.7 shows the result.

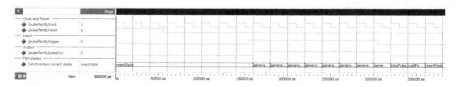

Fig. 4.7 Pulse FSM simulation in ModelSim that shows the eight clock cycle long pulse

4.3.1 Period-Doubling Route to Chaos

Example 4.3 Implement the Rössler period-doubling bifurcation from Sect. 4.2.1 on the FPGA.

Solution: Listing E.15 shows the VHDL specification. We have included the entire design specification instead of a snippet since we want the reader to understand the different steps involved in implementing bifurcations on the FPGA.

The DSP builder Advanced Blockset design that uses subsystems (hierarchical design) is shown in Figs. 4.8, 4.9, 4.10 and 4.11.

Figure 4.12 shows the result. Compare to Fig. 4.2. Note that the bifurcation parameter value(s) are approximately (since we are using 32-bit floating point) equal to those in Fig. 4.2.

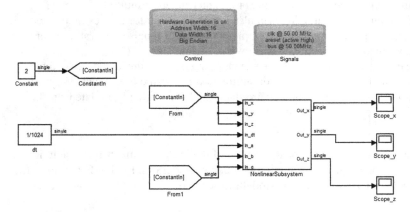

Fig. 4.8 Period-Doubling system top level

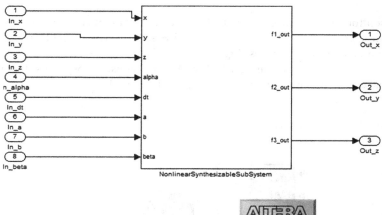

Fig. 4.9 Period-Doubling nonlinear subsystem

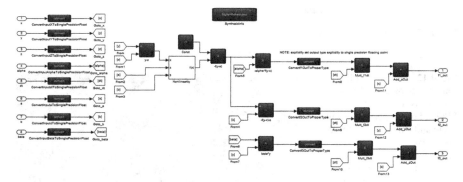

Fig. 4.10 Period-Doubling synthesizable subsystem

4.3.2 Period-Adding Route to Chaos

Example 4.4 Implement the Chua oscillator from Sect. 4.2.2 on the FPGA.

Solution: Listing E.16 shows the VHDL specification. We have again specified the full design because there are subtle differences between VHDL specifications of the different chaotic systems. It would be instructive to understand the reason for the differences. Nevertheless, we have not implemented the bifurcation mechanism in this example. Exercise 4.9 asks you to implement the period-adding route to chaos (Fig. 4.12).

The DSP builder Advanced Blockset design that uses subsystems (hierarchical design) is shown in Figs. 4.13, 4.14, 4.15 and 4.16.

Figure 4.17 shows the result.

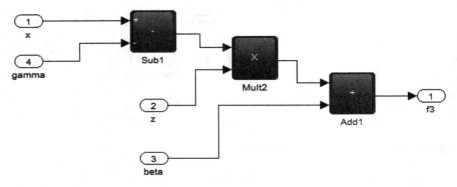

Fig. 4.11 Rössler nonlinearity

Fig. 4.12 Period-doubling route to chaos in the Rössler system, as realized on the FPGA. For the period-3 limit cycle and chaotic attractor, X and Y-axes scales are 0.5 V/div; for the period-6 limit cycle, the X-axis scale is 0.2 V/div and the Y-axis scale is 0.5 V/div

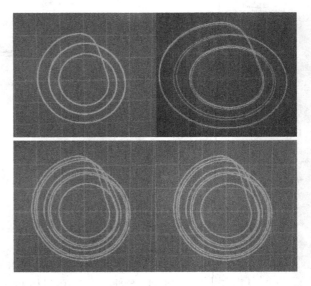

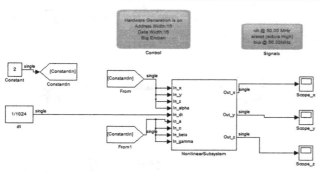

Fig. 4.13   Period-Adding system top level

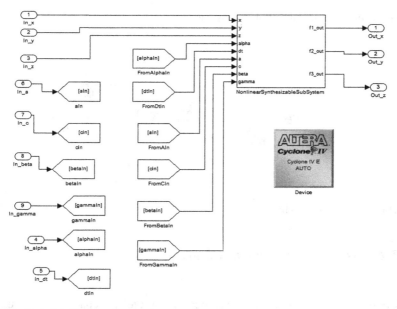

Fig. 4.14   Period-Adding nonlinear subsystem

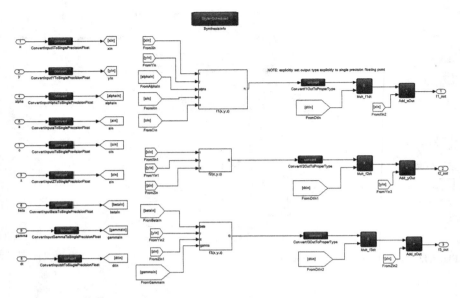

Fig. 4.15 Period-Adding synthesizable subsystem

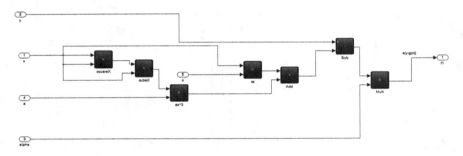

Fig. 4.16 Period-Adding nonlinearity

4.3.3 Quasi-Periodic Route to Chaos

Example 4.5 Implement the quasi-periodic route to chaos (via torus-breakdown) from Sect. 4.2.3 on the FPGA.

Solution: Listing E.17 shows the VHDL specification for the quasi-periodic route to chaos. We have again left the bifurcation implementation as an exercise for the reader in Problem 4.10.

Fig. 4.17 The Chua oscillator as realized on the FPGA. Scales are 0.2 V/div on both channels

The DSP builder Advanced Blockset design that uses subsystems (hierarchical design) is shown in Figs. 4.18, 4.19, 4.20 and 4.21.

Figure 4.22 shows the result.

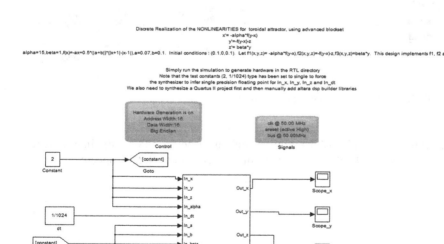

Fig. 4.18 Torus-breakdown system top level

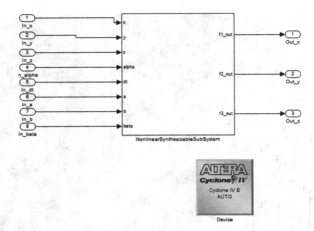

Fig. 4.19 Torus-breakdown nonlinear subsystem

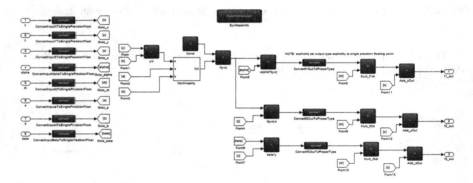

Fig. 4.20 Torus-breakdown synthesizable subsystem

Exercises 4.11 and 4.12 ask you to implement on the FPGA, the intermittency route to chaos and chaotic transients from Sects. 4.2.4 and 4.2.5 respectively.

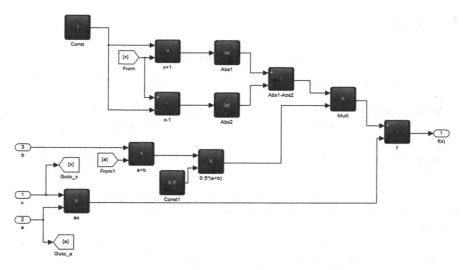

Fig. 4.21 Torus-breakdown nonlinearity

Fig. 4.22 Torus-breakdown route to chaos, as realized on the FPGA. Scales are 0.5 V/div on both channels

4.4 Conclusions

In this chapter, we studied bifurcations. In particular:

1. We understood bifurcations as a change in system behaviour when a parameter is varied.
2. We used the concept of bifurcations to study a variety of routes to chaos including period-doubling, period-adding, quasi-periodic (torus-breakdown), intermittency and chaotic transients.
3. We utilized single pulse generator to implement bifurcation mechanism on the FPGA.

This chapter has only scratched the surface of bifurcation phenomenon. Entire books have been written on this subject. The interested reader should pursue this topic further.

So far in this book, we have learned how to implement chaotic ODEs on an FPGA. In the concluding chapter to this volume, we will exercise the robustness of an FPGA by realizing chaotic DDEs. FPGA realization of DDEs is possible nowadays because of the copious amounts of on-chip memory.

Problems

4.1 Consider the Langford System, shown in Eqs. (4.26)–(4.28). These equations can be used to describe the motion of turbulent flow in a fluid [9].

$$\dot{x} = xz - \omega y \tag{4.26}$$

$$\dot{y} = \omega x + xy \tag{4.27}$$

$$\dot{z} = p + z - \frac{1}{3}z^3 - (x^2 + y^2)(1 + qx + \epsilon x) \tag{4.28}$$

First, compute the equilibrium points for the system. Now consider the following typical system parameters: $p = 1.1, q = 0.7$ and $\epsilon = 0.5$. Investigate the route to chaos in this system as a function of parameter ω. In other words, obtain a bifurcation diagram. Implement your design on the FPGA.

4.2 Investigate the route(s) to chaos in the Lorenz system.

4.3 Read [5] and obtain the period-doubling route to chaos in Eqs. (4.8)–(4.10)

4.4 Repeat Problem 4.3 but for the period-adding route.

4.5 Parameterize the pulseFSM in listing E.12 using generics. This will allow us to utilize the pulseFSM for other modules that require different delays.

4.6 We can also examine limit cycles with high periods (such as period-16) on the FPGA as opposed to an analog realization, due to noise immunity on the FPGA. Try to obtain high period limit cycles in any of the system(s) (say Rössler system) from this chapter.

4.7 Perform an In-system verification using SignalTap, of the period-doubling route to chaos for the Rössler system.

4.8 Investigate the route to chaos as a function of parameter κ, in the optically injected laser system in Eqs. (4.29)–(4.31) [10]. Use $\alpha = 2.5$, $\beta = 0.015$, $\Gamma = 0.05$, $\omega = 2$.

$$\dot{x}_1 = \kappa + \frac{x_1 x_3}{2} - \frac{\alpha}{2} x_2 x_3 + \omega x_2 \tag{4.29}$$

$$\dot{x}_2 = -\omega x_1 + \frac{\alpha}{2} x_1 x_3 + \frac{x_2 x_3}{2} \tag{4.30}$$

$$\dot{x}_3 = -2\Gamma x_3 - (1 + 2\beta x_3)(x_1^2 + x_2^2 - 1) \tag{4.31}$$

4.9 Following Example 4.3, implement the period-adding route to chaos Sect. 4.2.1.

4.10 Following Example 4.3, implement the quasi-periodic route to chaos Sect. 4.3.3.

4.11 Design and implement on the FPGA, the intermittency route to chaos from Sect. 4.2.4,

4.12 Design and implement on the FPGA, chaotic transients from Sect. 4.2.5.

Lab 4: Displaying Bifurcation Parameter(s) on the LCD

Objective: Implement a display module for bifurcation parameter(s).

Lab Exercise:

You should have realized from the design(s) in this chapter that simply pressing the key and mentally keeping track of the increment or decrement of the bifurcation parameter is cumbersome. Hence, utilize the solution from Lab 2 to display the current value of the bifurcation parameter on the LCD display. You can of course display any additional information or even interface to an external monitor using VGA.

References

1. Lakshmanan M, Rajasekar S (2003) Nonlinear dynamics—integrability, chaos and patterns. Springer, Berlin
2. Albers DJ, Sprott JC (2006) Routes to chaos in high-dimensional dynamical systems: a qualitative numerical study. Phys D 223:194–207
3. Chua LO, Wah CW, Huang A, Zhong G (1993) A universal circuit for studying and generating chaos—part I: routes to chaos. IEEE Trans Circuits Syst 40(10):732–744
4. Ambelang S (2011) Four routes to chaos: Chua's oscillator with a cubic nonlinearity. Final project report, EE4060 Spring 2011. Electrical Engineering and Computer Sciences Department, Milwaukee School of Engineering
5. Matsumoto T, Chua LO, Tokunaga R (1987) Chaos via torus breakdown. IEEE Trans Circuits Syst CAS 34(3):240–253
6. Pivka L, Wu CW, Huang A (1994) Chua's oscillator: a compendium of chaotic phenomena. J Frankl Inst 331(6):705–741
7. Kevorkian P (1993) Snapshots of dynamical evolution of attractors from Chua's oscillator. IEEE Trans Circuits Syst—I: Fundam Theory Appl 40(10):762–780
8. Dhamala M, Lai Y-C (1999) Controlling transient chaos in deterministic flows with applications to electrical power systems and ecology. Phys Rev E 59(2):1646–1655
9. Buscarino A et al (2014) A concise guide to chaotic electronic circuits. Springer, Berlin
10. Banerjee S, Saha P, Chowdhury AR (2004) Optically injected laser system: characterization of chaos, bifurcation and control. Chaos 14(2):347–357

Chapter 5
Chaotic DDEs: FPGA Examples and Synchronization Applications

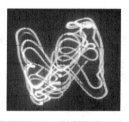

Valli et al. *Synchronization in Coupled Ikeda Delay Systems: Experimental Observations using FPGAs* [7]

Abstract This chapter explores particular advantage(s) of FPGAs for investigating nonlinear dynamics—realization of time delayed chaotic systems. These advantages are the availability of on-chip memory and the fact that generate statements in VHDL can be used to elegantly implement arbitrary (limited by on-chip memory and the number of FPGA logic elements) length delay chains. We will also explore synchronization applications in chaotic DDEs using the FPGA.

5.1 An Introduction to Time Delay Systems

Time delay is inherent in many physical systems and could be caused by (for example) lag between the sensing of disturbance and the triggering of an appropriate response [1, 4, 7]. Differential equations can be used to model time-delay systems and the general model that we will use in this chapter [5] is given in Eq. (5.1).

$$\dot{\mathbf{x}} = \mathbf{f}(t, \mathbf{x}(t), \mathbf{x}(t - \tau_i)) \tag{5.1}$$

In Eq. (5.1), $\mathbf{x} \overset{\triangle}{=} (x_1(t), x_2(t), \ldots, x_n(t))^T$ and $\tau_i > 0, i = 1, 2, \ldots, n$ are lag times or delay times. $\mathbf{f}$ is a vector valued continuous function.

© Springer International Publishing Switzerland 2015

B. Muthuswamy and S. Banerjee, *A Route to Chaos Using FPGAs*, Emergence, Complexity and Computation 16, DOI 10.1007/978-3-319-18105-9_5

We will utilize the forward-Euler's method (recall Sect. 3.1) for specifying the DDE, refer to Eq. (5.2).

$$\mathbf{x}(t + \delta t) = \mathbf{x}(t) + \mathbf{f}(\mathbf{x}(t), \mathbf{x}(t - N_i))\Delta t \tag{5.2}$$

In Eq. (5.2), we have slightly abused our notation and have used continuous time t whereas the equation is actually discrete. Nevertheless, the advantage of an FPGA becomes apparent when realizing the delay(s) N_i. But let us first simulate DDEs in Simulink.

5.2 Simulating DDEs in Simulink

A general simulation block diagram is shown in Fig. 5.1. Example 5.1 shows how to adapt the block diagram in Fig. 5.1 to a particular chaotic DDE.

Example 5.1 Simulate the Ikeda DDE [3] in Simulink.

$$\dot{x} = \mu \sin(x(t - \tau)) - \alpha x(t) \tag{5.3}$$

$\mu = 6, \tau = 1, \alpha = 1$.

Solution: Fig. 5.2 shows the Ikeda DDE simulation block diagram. Phase plot is shown in Fig. 5.3.

Figure 5.1 implies that DDEs can be realized on an FPGA quite easily, once the delay has been specified. The subject of FPGA realization is the topic of the next section.

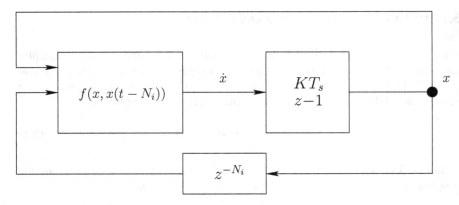

Fig. 5.1 Simulating a DDE in simulink, block diagram adopted from [7]. We utilize the discrete-time integrator but inherit the sample time from the fixed-point simulation period

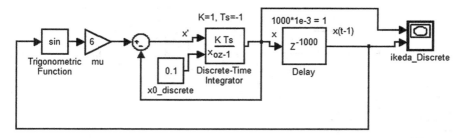

Fig. 5.2 Simulink block diagram for simulating the Ikeda DDE. Fixed-step Euler's method was used with a step size 0.001

Fig. 5.3 Ikeda DDE simulation result. y-axis is $x(t-1)$, x-axis is $x(t)$

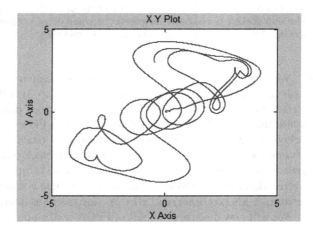

5.3 FPGA Realization of DDEs

In order to implement the delay on an FPGA, we will need the definition in Eq. (5.4) [5].

$$\Delta t \overset{\triangle}{=} \frac{\tau_i}{N_i - 1} \tag{5.4}$$

Based on Eq. (5.4), we can utilize the for statement (refer to listing C.1) to let the synthesizer infer the number of flip-flops required for the delay.

Line 22 in listing F.2 declares the internal delay lines to be 32 bits wide. However, we need an array of these internal delay lines and hence we have declared it as "memory". Thus, we are specifying a tapped delay line in VHDL.

Figure 5.4 is the block diagram for FPGA realization of DDEs.

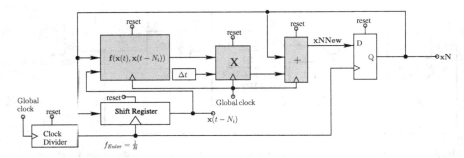

Fig. 5.4 Hardware block diagram for specifying DDEs, using forward-Euler method [7]. Note that one could specify more advanced integration methods such as Runge-Kutta. The block diagram utilizes a combination of DSP builder advanced block set (blocks highlighted in *grey*) and VHDL. The shift register is obviously unnecessary for non-delay systems and thus the block diagram specifies how we can implement nonlinear ODEs on FPGAs

We can infer from Fig. 5.4 that the only modification to our already existing FPGA realization of chaotic systems (starting from Chap. 3) is the delay realization using VHDL from listing F.1 and F.2. In Fig. 5.4, the clock divider block is configured to divide the global clock (usually obtained from the FPGA board clock) so propagation delays associated with the various sub-modules (such as the shift register) can be accommodated. Hence the clock divider output clock (with frequency $f_{Euler} = \frac{1}{\delta t}$) is used as clock input for the shift register and the D flip-flop synchronizer. Since the overall design is synchronous, all sequential logic components have a well-defined reset state. $\mathbf{xN}$ ($\mathbf{x}(t + \delta t)$) and $\mathbf{x}(t - N_i)$ also serve as inputs to the audio codec DAC.

We will now discuss examples of FPGA realization of DDEs.

Example 5.2 Implement the Ikeda DDE from Example 5.2 on the DE2.

Solution: In order to implement the Ikeda DDE, we will simply utilize the D flip-flop (with async reset) and the addressable shift register from listings F.1 and F.2 resp. Listing F.3 shows the complete VHDL Ikeda module.

Please refer to the online video and reference design on the companion website: http://www.harpgroup.org/muthuswamy/ARouteToChaosUsingFPGAs/ReferenceDesigns/volumeI-ExperimentalObservations/chapter5/ikedaDDE/ for further details on the Ikeda DDE. Nevertheless, we have shown all the DSP builder subsystems (similar to Chap. 4) in Figs. 5.5, 5.6, 5.7 and 5.8.

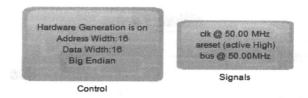

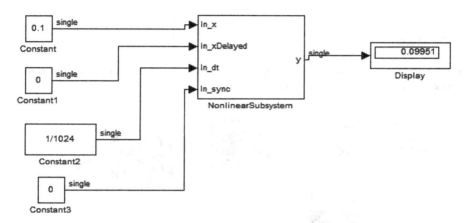

Fig. 5.5 Ikeda system top level

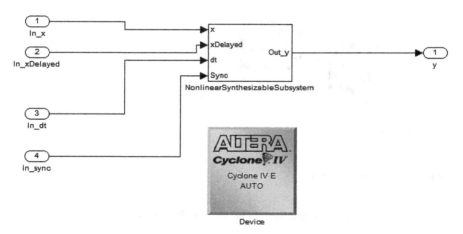

Fig. 5.6 Ikeda nonlinear subsystem

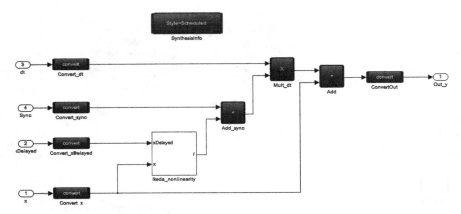

Fig. 5.7 Ikeda synthesizable subsystem. The optional sync input can be set to zero if we are not performing synchronization experiments

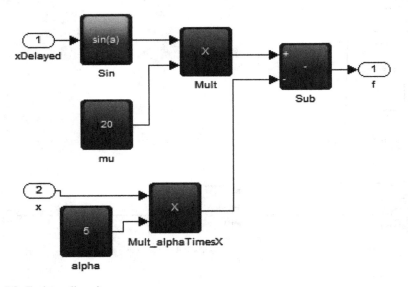

Fig. 5.8 Ikeda nonlinearity

Figure 5.9 shows the result.

Example 5.3 Implement the Sigmoidal DDE in Eq. (5.5) on the FPGA.

$$\dot{x} = 2\tanh(x(t - \tau)) - x(t - \tau) \tag{5.5}$$

Use $\tau = 3$.

Solution: Although DSP builder advanced blockset does not have a hyperbolic tangent function, recall from Sect. 3.7 that the hyperbolic tangent can be written in terms of exponential functions, refer to Eq. (5.6).

$$\tanh(x) = \frac{e^x - e^{-x}}{e^x + e^{-x}} \tag{5.6}$$

We can thus implement the sigmoidal DDE since the exponential function is available in DSP builder advanced blockset. Listing F.4 shows the complete VHDL specification for the sigmoidal DDE.

Figures 5.10, 5.11, 5.12 and 5.13 show the DSP builder design. Figure 5.14 shows the result.

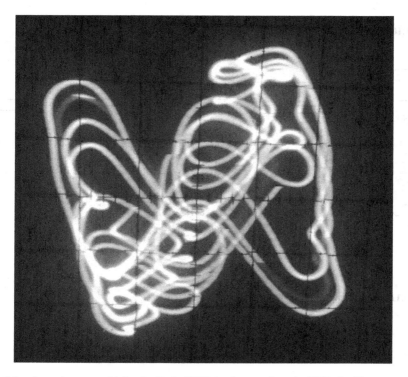

Fig. 5.9 $x(t - \tau)$ verses $x(t)$ for the Ikeda DDE, implemented on the DE2. Oscilloscope scales are 0.5 V/div for both axis

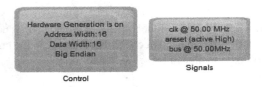

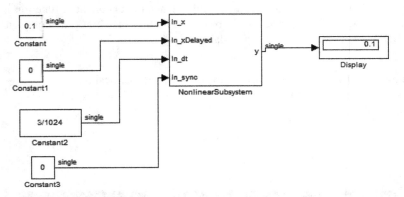

Fig. 5.10 Sigmoid DDE DSP builder system top level

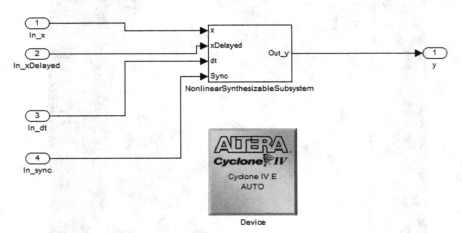

Fig. 5.11 Sigmoid DDE DSP builder nonlinear subsystem

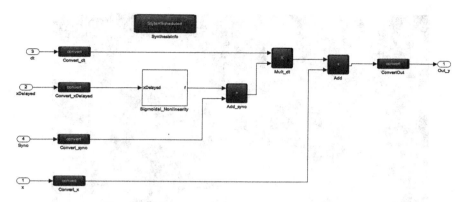

Fig. 5.12 Sigmoid DDE DSP builder synthesizable subsystem

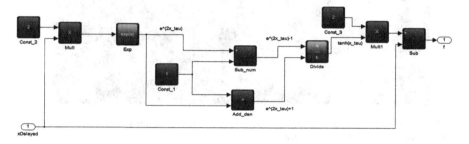

Fig. 5.13 Sigmoid DSP builder nonlinearity

Example 5.4 Implement the Signum DDE in Eq. (5.7) on the FPGA.

$$\dot{x} = \text{sgn}(x(t - \tau)) - x(t - \tau) \qquad (5.7)$$

Use $\tau = 2$.

Solution: Listing F.5 shows the complete VHDL specification for the signum DDE. Note that we do not use DSP builder advanced block set for the nonlinearity, since it is so trivial to implement.

Figure 5.15 shows the result.

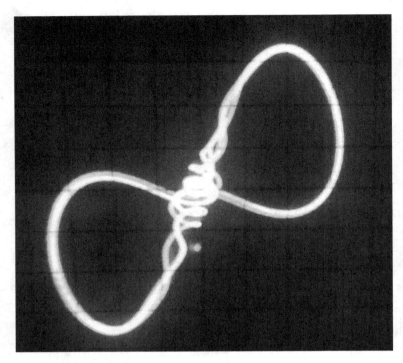

Fig. 5.14 $x(t - \tau)$ verses $x(t)$ for the sigmoid DDE, implemented on the DE2. Oscilloscope scales are 0.1 V/div for X-axis and 1 V/div for Y-axis

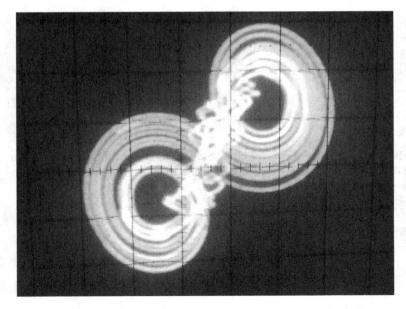

Fig. 5.15 $x(t - \tau)$ verses $x(t)$ for the signum DDE, implemented on the DE2. Oscilloscope scales are 0.1 V/div for X-axis and 1 V/div for Y-axis

5.4 Applications of (Time Delayed) Chaotic Systems—Synchronization

In Sect. 1.1.2, we briefly touched upon the topic of synchronization. In this section, we will expand upon this topic further. We will briefly discuss the concept of synchronization first and then show examples of synchronization in chaotic DDEs. One of the reasons why we focus on DDEs is because they are infinite dimensional [5]. Hence they are more attractive to applications such as secure communications, compared to chaotic systems without delay [5].

A surprising property of chaotic attractors is their susceptibility to synchronization [8]. This refers to the tendency of two or more systems that are coupled together to undergo closely related motions, even if the motions are chaotic. This property is surprising because it was believed that chaotic synchronization was not feasible because of the hallmark property of chaos: sensitive dependence on initial conditions [5]. Hence chaotic systems intrinsically defy synchronization because even two identical systems starting from very slightly different initial conditions would evolve in time in an unsynchronized manner (the differences in state would grow exponentially). Nevertheless it has been shown that it is possible to synchronize chaotic systems [5, 8], to make them evolve on the same trajectory, by introducing appropriate coupling between them due to the works of Pecora and Carroll and the earlier works of Fujisaka and Yamada [5].

Chaos synchronization has been receiving a great deal of interest for more than two decades in view of its potential applications in various fields of science and engineering [5]. There are a variety of synchronization mechanisms that have been proposed: complete or identical synchronization, phase synchronization, almost synchronization, episodic synchronization—are a few [5]. A detailed discussion of synchronization mechanisms is obviously beyond the scope of this chapter or this book. We will however give an example of chaotic DDE synchronization on FPGAs using the Ikeda DDE [7].

Complete synchronization is the simplest type of synchronization that is characterized by perfect follow-up of two chaotic trajectories. Synchronization is achieved by means of a coupling function. We consider linearly coupled Ikeda systems as the drive and response systems, described by Eqs. (5.8) and (5.9) respectively.

$$\dot{x} = -\alpha x + \mu \sin x(t - \tau) \tag{5.8}$$

$$\dot{y} = -\alpha y + \mu \sin y(t - \tau) + k(t)(x - y) \tag{5.9}$$

$k(t)$ is the coupling function between drive and response system. In the following subsections, we illustrate complete synchronization using unidirectional and bidirectional coupling via the FPGA. The parameters are taken as $\mu = 20, \alpha = 5, \tau = 1$. Figure 5.16 shows the analog output from Eq. (5.3), with the parameters for synchronization.

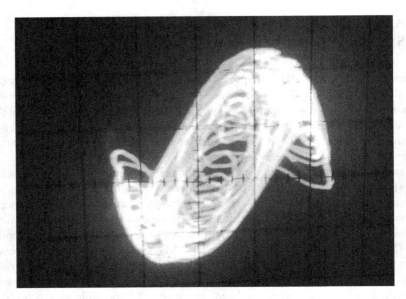

Fig. 5.16 $x(t - \tau)$ (Y-input) verses $x(t)$ (X-input) displayed on an oscilloscope for the Ikeda attractor with parameters for synchronization experiments. Scales for both channels are 0.2 V/div

5.4.1 Unidirectional Coupling

In unidirectional coupling, the drive system is free to evolve and the response system is influenced by the coupling function $k(t)$. Due to this, the dynamical evolution of the response system is to follow the dynamics of the drive system in the course of time. We investigated chaos synchronization in unidirectionally coupled Ikeda systems with two types of coupling functions. In the first type, $k(t)$ is a square wave coupling represented by

$$(t_0, k_1), (t_1, k_2), (t_2, k_1), (t_3, k_2)....... \tag{5.10}$$

where $t_j = t_0 + (j - 1)\tau$, $j \geq 1$ with $k_1 \neq k_2$. It is observed that the amplitude of the control parameter $k(t)$ is the key factor to achieve synchronization between drive and response, larger the amplitude quicker the convergence into synchronization, provided that the conditional Lyapunov exponents of the response systems are all negative. The threshold value is chosen as $k_1 = 0$ and $k_2 = 50$ for square wave coupling. We also used the second type of coupling function defined in Eq. (5.11).

$$k(t) = -\alpha + 2\mu |\cos(y(t - \tau))| \tag{5.11}$$

5.4.2 Bidirectional Coupling

In bidirectional coupling, both drive and response systems are coupled with each other by a coupling function $k(t)$ that induces mutual synchronization. For this bidirectional coupling, the drive Eq. (5.12) and response Eq. (5.13) systems are considered as

$$\dot{x} = -\alpha x + \mu \sin x(t - \tau) + k(t)(y - x) \tag{5.12}$$

$$\dot{y} = -\alpha y + \mu \sin y(t - \tau) + k(t)(x - y) \tag{5.13}$$

The synchronization error is computed by $e(t) = x(t) - y(t)$, which is the measure for convergence of two chaotic trajectories. Figures 5.17, 5.18, 5.19 and 5.20 show results from the analog output synchronization experiment. In order to generate the

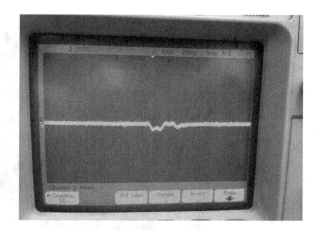

Fig. 5.17 Synchronization error for unidirectional square wave coupling. Vertical scale is 200 mV/div., horizontal scale is 200 μs/div

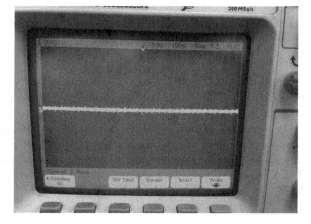

Fig. 5.18 Synchronization error for unidirectional cosine function based coupling. Vertical scale is 200 mV/div., horizontal scale is 100 μs/div

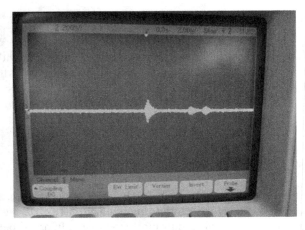

Fig. 5.19 Synchronization error for bidirectional square wave coupling. Vertical scale is 200 mV/div., horizontal scale is 2.00 μs/div

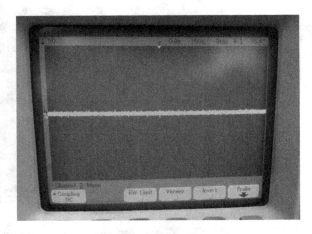

Fig. 5.20 Synchronization error for bidirectional cosine function based coupling. Vertical scale is 200 mV/div., horizontal scale is 100 μs/div

analog output, we simply passed $x(t) - y(t)$ into a DAC channel. It is observed that the synchronization is quicker in bidirectional coupling compared to the unidirectional coupling for the same parameters.

In order to shed more light on synchronization, we have used the analog oscilloscope to obtain X-Y plots where X is $x(t)$ (drive) and Y is $y(t)$ (response). Figure 5.21 first shows the X-Y plot when the drive and response systems are not synchronized. The X-Y plots as a result of synchronization are shown in Figs. 5.22, 5.23, 5.24 and 5.25.

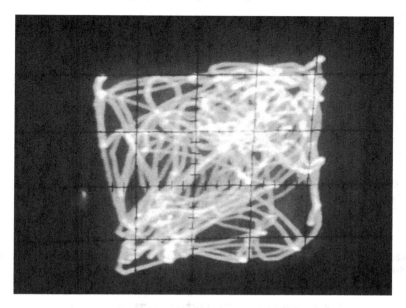

Fig. 5.21 The fact that the drive and response systems are not synchronized and the outputs are analog waveforms is an experimental evidence of "sensitive dependence on initial conditions", one of the hallmarks of chaotic systems. Although the FPGA Ikeda DDE drive and response systems are both digital specifications, the output waveform is analog. Thus any noise on the analog line is going to ensure that the drive and response systems do not have the same initial conditions and this leads to the X-Y plot shown

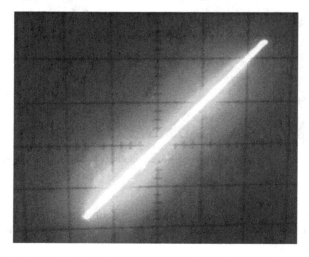

Fig. 5.22 XY plot for unidirectional square wave coupling. Vertical and horizontal scales are 200 mV/div

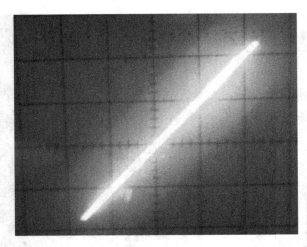

Fig. 5.23 XY plot for unidirectional cosine function based coupling. Vertical and horizontal scales are 200 mV/div

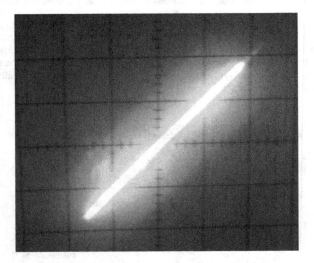

Fig. 5.24 XY plot for bidirectional square wave coupling. Vertical and horizontal scales are 200 mV/div

Listing F.6 shows one possible approach for implementing synchronization schemes on an FPGA. The top level is shown in listing F.7 so that we may fully understand the design.

Fig. 5.25 XY plot for bidirectional cosine function based coupling. Vertical and horizontal scales are 200 mV/div

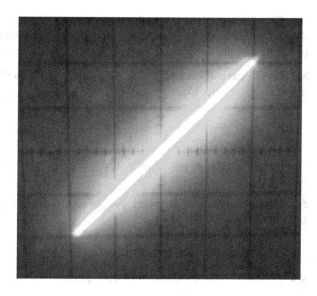

5.5 Conclusions

In this chapter, we understood that chaotic DDEs can be realized on FPGAs using tapped delay lines. We also studied an application of chaotic systems (on FPGAs) to synchronization. DDEs are particularly suited for synchronization and secure communication applications because of their infinite dimensionality.

Applications aside, we hope you had fun understanding that FPGAs are robust physical platforms for realizing nonlinear (chaotic) ODEs. In a followup volume, we will examine theoretical methods to rigorously understand the implications of sampling and quantization on the underlying system behaviour.

Acknowledgments Many thanks to our colleagues at the Vellore Institute of Technology, Vellore, India for working with us on the synchronization experiments. Specifically, Ph.D. candidate Ms. Valli, Professors Ganesan and Subramanian have been extremely helpful.

Problems

5.1 Consider the Ikeda DDE:

$$\dot{x} = \mu \sin(x(t - \tau)) - \alpha x(t) \qquad (5.14)$$

$\mu = 6, \tau = 1, \alpha = 1$.
Perform a ModelSim simulation of the system above.

5.2 One approach to speed up the synthesis procedure is to minimize the number of delays by increasing the sampling frequency of the system. Explore this approach by increasing the sampling frequency for, say, the Ikeda system.

5.3 One of the earliest and most widely studied DDE is the Mackey-Glass equation [6], shown in Eq. (5.15).

$$\dot{x} = \frac{ax(t-\tau)}{1+x(t-\tau)^c} - bx(t) \tag{5.15}$$

Parameters for chaos: $a = 3, b = 1, c = 7, \tau = 3$ [6]. Implement the equation on the FPGA.

5.4 Implement the antisymmetric piecewise-linear DDE [6], shown in Eq. (5.16) on the FPGA. Use $\tau = 3$.

$$\dot{x} = |x(t-\tau)+1| - |x(t-\tau)-1| - x(t-\tau) \tag{5.16}$$

5.5 Implement the asymmetric piecewise-linear DDE [6], shown in Eq. (5.17) on the FPGA. Use $\tau = 1.8$.

$$\dot{x} = x(t-\tau) - 2|x(t-\tau)| + 1 \tag{5.17}$$

5.6 Explore the synchronization schemes discussed in Sect. 5.4 using the DDEs from problems 5.3, 5.4 and 5.5.

5.7 Consider Eq. (5.18) [6].

$$\dot{x} = \frac{1}{\tau} \int_0^\tau x(t-s)(4-|x(t-s)|)ds \tag{5.18}$$

In Eq. (5.18), the time derivative depends on the average value of a function for time lags of x_s from $s = 0$ to τ. Implement the equation on an FPGA, using $\tau = 3$.

5.8 Investigate bifurcation mechanisms in any of the DDEs from this chapter.

Lab 5: The Lang-Kobayashi Chaotic Delay Differential Equation

Objective: Simulate and physically realize the Lang-Kobayashi (L-K) chaotic DDE [2]

$$\frac{dE}{dt} = -(1+i\alpha)|E|^2E + \eta_1 E(t-\tau_1) + \eta_2 E(t-\tau_2) \tag{5.19}$$

Theory: Notice that Eq. (5.19) is in the complex domain. However we can separate the real and imaginary parts by writing $E(t) = \rho(t)e^{i\theta(t)}$ in Eq. (5.19) to obtain Eqs. (5.20) and (5.21).

$$\frac{d\rho}{dt} = -\rho^3 + \eta_1 \rho(t - \tau_1) \cos(\theta(t) - \theta(t - \tau_1))$$
$$+ \eta_2 \rho(t - \tau_2) \cos(\theta(t) - \theta(t - \tau_2)) \tag{5.20}$$

$$\rho\frac{d\theta}{dt} = -\alpha\rho^3 + \eta_1 \rho(t - \tau_1) \sin(\theta(t) - \theta(t - \tau_1))$$
$$+ \eta_2 \rho(t - \tau_2) \sin(\theta(t) - \theta(t - \tau_2)) \tag{5.21}$$

Verify that one can indeed obtain Eqs. (5.20) and (5.21) from Eq. (5.19).

Lab Exercise:

1. Simulate (using Simulink and Modelsim), verify using SignalTap and hence implement Eqs. (5.20) and (5.21) for the following parameters $\alpha = 4, \eta_1 = 3.5, \eta_2 = 3, \tau_1 = 2.5, \tau_2 = 0.1$.
2. Study synchronization mechanisms in the L-K DDE using the ideas from Sect. 5.4.

References

1. Banerjee S, Rondoni L, Mukhopadhyay S (2011) Synchronization of time delayed semiconductor lasers and its applications in digital cryptography. Opt Commun 284:4623–4634
2. Banerjee S, Ariffin MRK (2013) Noise induced synchronization of time-delayed semiconductor lasers and authentication based asymmetric encryption. Opt Laser Technol 45:435–442
3. Ikeda K, Daido H, Akimoto O (1980) Optical turbulence: chaotic behavior of transmitted light from a ring cavity. Phys Rev Lett 45:709
4. Jeeva STS, Ariffin MRK, Banerjee S (2013) Synchronization and a secure communication scheme using optical star network. Opt Laser Technol 54:15
5. Lakshmanan M, Senthilkumar DV (2011) Dynamics of nonlinear time-delay systems. Springer Series in Synergetics, New York
6. Sprott, JC (2010) Elegant chaos. World Scientific
7. Valli D et al (2014) Synchronization in coupled Ikeda delay differential equations: experimental observations using field programmable gate arrays. Eur Phys J Spec Top 223(8):1–15. doi:10.1140/epjst/e2014-02144-8
8. Wolfson (2013) WM8731 datasheet, Available via DIALOG. http://www.wolfsonmicro.com/products/audio_hubs/WM8731/ Accessed 4 Oct 2013

Appendix A
Introduction to MATLAB and Simulink

In this appendix, we will discuss MATLAB and Simulink. Note that we will only cover aspects of MATLAB and Simulink that are useful for simulating nonlinear differential equations. MATLAB is an acronym for MATrix LABoratory and is a product of the MathWorks corporation [1]. Simulink is a graphical front end to MATLAB. This appendix assumes you have MATLAB version 7.1 (R14) installed (with default options). Note that any version of MATLAB later than 7.1 should work although there may be slight differences between your version of MATLAB and the version used in this appendix.

Note our approach is not the only procedure for simulating differential equations. The reader is encouraged to independently explore other methods on their own.

A.1 Simulating Nonlinear Differential Equations in MATLAB

A.1.1 Starting MATLAB

First, start MATLAB by left-clicking on the MATLAB icon ▰. Figure A.1 should appear. Some of the windows maybe disabled by default. Make sure that **Window** → **Command History** and **Window** → **Workspace** are enabled. These windows enable you to quickly retype a previous command and check MATLAB's memory usage respectively.

The online help system in MATLAB is extensive. If you are unfamiliar with MATLAB, you should understand the basics of MATLAB before proceeding further. Press "F1" on your keyboard to bring up the MATLAB help window, a portion of which is shown in Fig. A.2. You should go through the "Getting Started" section and as many sections as necessary to understand MATLAB.

© Springer International Publishing Switzerland 2015   123
B. Muthuswamy and S. Banerjee, *A Route to Chaos Using FPGAs*, Emergence,
Complexity and Computation 16, DOI 10.1007/978-3-319-18105-9

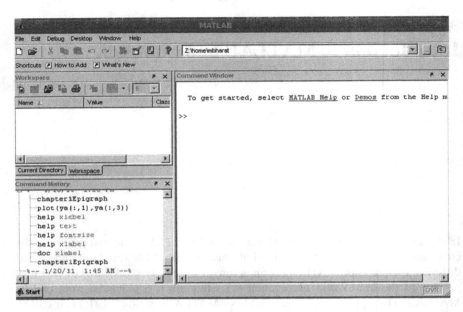

Fig. A.1 The startup screen in MATLAB, the command prompt is indicated by "≫"

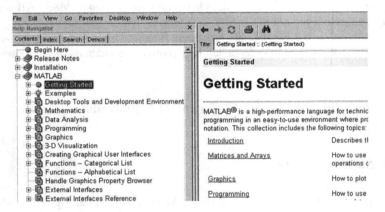

Fig. A.2 The MATLAB help window

A.1.2 Simulating a One Dimensional System in MATLAB

Consider the following one dimensional dynamical system

$$\dot{x} = \sin x, \ x(0) = \frac{\pi}{2} \tag{A.1}$$

The code required to simulate the system is shown below. The MATLAB comments should make the code self-explanatory. Note that you should liberally comment your code to improve readability. While labeling the plot of our solution, we have used MATLAB's ability to interpret LATEX commands for improved readability. For more information, type "help latex" at the MATLAB command prompt.

```
1    % Lines starting with a % are comment.
2    % MATLAB code for simulating a one dimensional nonlinear
3    % dynamical system
4    % Muthuswamy, Bharathwaj
5
6    % The lines below instruct MATLAB to clear all workspace variables
7    % (or memory). It is a good idea to start your simulation from a
8    % clean MATLAB state in order to eliminate the side-effects caused
9    % by unused variables. The semicolon at the end of a line
10   % supresses echo to the MATLAB command line.
11   clear all;
12   close all;
13
14   % The line below defines our system via the "inline" MATLAB
15   % command. The first argument is our system. The second
16   % argument defines the independent variable (time) and the
17   % third argument defines the array "y" for our dependent
18   % variable. Thus y(1) = x(t).
19   sinusoidalSystem = inline('[sin(y(1))]','t','y');
20
21   % We setup tolerance options for the Ordinary Differential
22   % Equation (ODE) solver.
23   % The values below suffice for our systems.
24   options = odeset('RelTol',1e-7,'AbsTol',1e-7);
25
26   % The line below invokes the medium order ode45 solver,
27   % we will use this solver for our systems. The first argument
28   % is the system to be solved.
29   % The second argument is a matrix with start and stop times.
30   % The third argument specifies the initial conditions and
31   % the fourth argument uses the options specified previously.
32   [t,ya] = ode45(sinusoidalSystem,[0,100],[pi/2],options);
33
34   % plot the solution. The first argument to the plot
35   % command is the x-axis variable and the second argument
36   % is the y-axis variable.
37   plot(t,ya(:,1));
38
39   % Label axis with units. Then title the plot.
40   % Note that we use a Latex interpreter.
41   xlabel('$t$ (seconds)','FontSize',14,'Interpreter','Latex');
42   ylabel('$x(t)$','FontSize',14,'Interpreter','Latex');
43   title('Solution of $\dot{x} = \sin(x), x(0)=\frac{\pi}{2}$','Interpreter','Latex');
```

The reader is encouraged to use script files (or M-files in MATLAB terminology) to enter your commands so you can save them for reuse in a later MATLAB session. To create an M-file and enter the commands above, go to **File → New → M − File**. Enter the commands above and save the file as "oneDimensionalSimulation.m". The result is shown in Fig. A.3. You can run the file by typing the filename in the MATLAB command prompt and pressing enter.

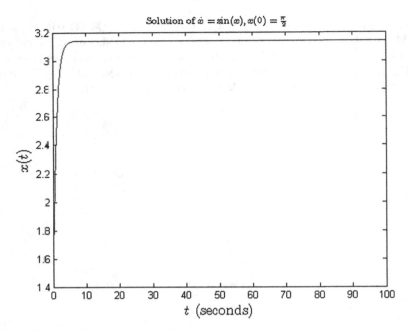

Fig. A.3 The result of simulating our one dimensional system

We have already seen how to simulate a chaotic system in MATLAB (the Lorenz system in Chap. 1). Let us now understand how to use Simulink, the graphical front-end to MATLAB.

A.2 Simulating Nonlinear Differential Equations in Simulink

In this section, we will show you how to simulate nonlinear differential equations using Simulink. This tool offers a more visual approach to the differential equation setup.

A.2.1 Starting Simulink

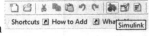

First, start Simulink by left-clicking on the Simulink icon in the MATLAB tool bar. Figure A.4 should pop up.

The Simulink library browser contains a plethora of components. We will restrict ourselves to the **Integrator** block (highlighted in Fig. A.4) under the **Continuous** Library and various blocks in the **Math Operations**, **User-Defined Functions**, **Sinks** and **Sources** library.

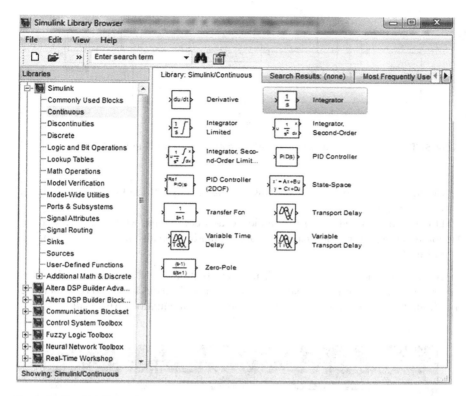

Fig. A.4 Simulink library browser

A.2.2 Simulating a One Dimensional System in Simulink

We will simulate the system shown below.

$$\dot{x} = e^{-x} - \cos(x), \quad x(0) = 0.1 \tag{A.2}$$

First open a new model in Simulink by left-clicking **File** $\rightarrow$ **New** $\rightarrow$ **Model**

Figure A.5 shows the system represented in Simulink. In order to construct this system, follow the steps below.

1. Place an **Integrator** block from the **Continuous** Library by left-clicking the block and dragging it into your model. Double-click the block and set the initial condition to 0.1.
2. Next place a **Fcn** block from the **User-Defined Functions** library.
3. Double-click the **Fcn** block and enter "exp(-u)-cos(u)". Notice that "u" is the input variable.
4. Drag and place a **Scope** block from the **Sinks** library.

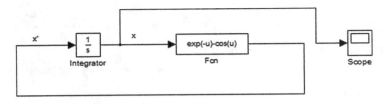

Simulating x' = e^(-x) - cos(x), x(0) = 0.1

Fig. A.5 The one dimensional system $\dot{x} = e^{-x} - \cos(x),\; x(0) = 0.1$ in Simulink

5. Connect all the components as shown in Fig. A.5 by left-clicking and dragging a wire connection.
6. Double-click anywhere in the model to add comments. Make sure you add a comment indicating the system you are simulating and also label wires, as shown in Fig. A.5.

To simulate the system, left-click the Play button ◁◀▷▷▯▷▶ᴵᴼᴼ in the Simulink toolbar. The default options are sufficient for the models in this book. You can increase the simulation time appropriately, for this differential equation, 10 s is sufficient.

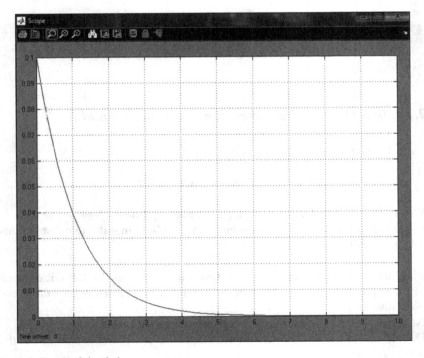

Fig. A.6 Result of simulating our system

The result should be Fig. A.6. Note that you cannot unfortunately name the axes and title the plot. The colors have been inverted for printing purposes.

A.2.3 Simulating a Chaotic System in Simulink

Now we will simulate the Sprott system shown below.

$$\dddot{x} + \ddot{x} + x + f(\dot{x}) = 0 \tag{A.3}$$

The nonlinear function is given by:

$$f(\dot{x}) = \mathrm{sign}(1 + 4\dot{x}) \tag{A.4}$$

Here $\mathrm{sign}(x)$ is the signum function given by:

$$\mathrm{sign}(x) = \begin{cases} -1 & \text{when } x < 0, \\ 0 & \text{when } x = 0, \\ 1 & \text{when } x > 0. \end{cases}$$

The Simulink model of our system is shown in Fig. A.7. Simulating our system for 100 s, the result should be Fig. A.8.

The **Sum** block can be found in **Math Operations** library. The **Sprott Nonlinear System** is an **XY Scope** from the **Sinks** library. The initial conditions are $x(0) = -0.5$, $y(0) = z(0) = 0$. *sgn* is used with the **Fcn** block under the **User-Defined Functions** library.

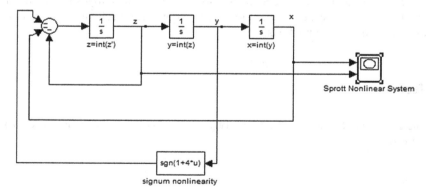

Fig. A.7 Simulating the Sprott system in Simulink

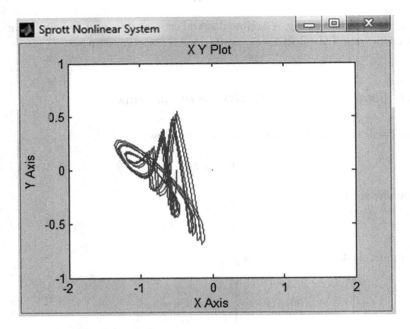

Fig. A.8 Results of simulating the Sprott system

A.3 Conclusion

In this appendix we showed you how to simulate nonlinear differential equations in MATLAB and Simulink.

The reader may have noticed that we are simulating a sensitive system on a finite state machine (computer). How can we be even confident that our simulation is correct? It turns out that the concept of "shadowing" can be used to justify numerical simulation of chaotic systems.[1] For more information, please refer to the upcoming volume II on theoretical methods.

Reference

1. The Mathworks Corporation (2012) Available via DIALOG. http://www.mathworks.com. Accessed 25 Dec 2012

---

[1]We cannot invoke the Nyquist-Shannon sampling theorem to determine a sampling frequency since our system is not bandlimited.

Appendix B
Chapter 1 MATLAB Code

B.1 The Lorenz System

Listing B.1 MATLAB code for Lorenz equations

```
1   % Lines starting with % are comments in MATLAB.
2   % Extensively comment your code!
3   % Purpose of this code: Obtaining phase plots and time-domain
4   % waveforms for Lorenz system.
5   %
6   % The lines below instruct MATLAB to clear all workspace
7   % variables (or memory). It is a good idea to start your
8   % simulation from a clean MATLAB state in order to eliminate
9   % the side-effects caused by unused variables. The semicolon
10  % at the end of a line will supress echo to the MATLAB command
11  % line.
12  clear all;
13  close all;
14
15  % The line below defines our system via the "inline" MATLAB
16  % command. The first argument is our system. Since our system
17  % is three dimensional, we have a 3 x 1 matrix for our system.
18  % The second argument defines the independent variable (time)
19  % and the third argument defines the array "y" for our
        dependent
20  % variable. Thus y(1) = x(t),y(2) = y(t) and y(3) = z(t).
21  lorenz = inline('[-10*y(1)+10*y(2);-y(1)*y(3)+28*y(1)-y(2);y
        (1)*y(2)-(8/3)*y(3)]','t','y');
22
23  % We setup tolerance options for the Ordinary Differential
24  % Equation (ODE) solver. The values below suffice for our
25  % systems.
26  options = odeset('RelTol',1e-7,'AbsTol',1e-7);
27
28  % The line below invokes the medium order ode45 solver, we
        will
```

```
29  % use this solver for our systems. The first argument is the
30  % system to be solved. The second argument is a matrix with
31  % start and stop times. The third argument specifies the
        initial
32  % conditions and the fourth argument uses the options
        specified
33  % in the line above.
34  % You should use initial conditions very close to the
        equilibrium
35  % points.
36  h1=figure;
37  [t,ya] = ode45(lorenz,[0,100],[10,20,30],options);
38
39  % classic view of lorenz butterfly
40  % the first argument is the x-value to be plotted
41  % the second argument is the y-value
42  plot(ya(:,1),ya(:,3));
43  % clearly label your axes!
44  xlabel('$x$','Interpreter','LaTeX','FontSize',32);
45  ylabel('$z$','Interpreter','LaTeX','FontSize',32);
46  % uncomment line below for EPS output
47  % print(h1,'-depsc','-r600','chap1Figure-LorenzAttractor2D.eps
        ');
48  % new figure for three dimensional plot
49  h2 = figure;
50  % plot3 is very similar to plot
51  plot3(ya(:,1),ya(:,2),ya(:,3));
52  % the view point setting below was experimentally determined
        to
53  % see the best possible view of the attractor, in order to
54  % understand the sheet-like nature of the fractal. You should
55  % rotate and zoom the view in the 3D MATLAB figure to better
56  % understand the structure.
57  azimuth = -76; % degrees
58  elevation = -40; % degrees
59  view(azimuth,elevation);
60  xlabel('$x$','Interpreter','LaTeX','FontSize',24);
61  ylabel('$y$','Interpreter','LaTeX','FontSize',24);
62  zlabel('$z$','Interpreter','LaTeX','FontSize',24);
63  % uncomment line below for EPS output
64  % print(h2,'-depsc','-r600','chap1Figure-LorenzAttractor3D.eps
        ');
65  % time domain plot, not visually appealing :)
66  h3 = figure;
67  % we plot only 5000 points so we can see some features of
68  % the waveform clearly
69  % remember: MATLAB array indices start at 1.
70  plot(t(1:5000),ya([1:5000],1));
71  xlabel('$t$ (seconds)','Interpreter','LaTeX','FontSize',24);
72  ylabel('$x$','Interpreter','LaTeX','FontSize',24);
73  % uncomment line below for EPS output
```

```
74   % print(h3,'-depsc','-r600','chap1Figure-
         LorenzAttractorTimeDomain.eps');
```

B.2 Linear Equation

Listing B.2 MATLAB code for plotting the graph of a straight line

```
1    % Purpose: plot simple linear system
2    % define range of x values (-5 to 5) and spacing (0.1) between
3    % values.
4    close all;
5    clear all;
6    x = [-5:0.1:5];
7    % define function and plot the resulting straight line.
8    y=(x/3)-1;
9    h1=figure;
10   plot(x,y)
11   xlabel('$x$','Interpreter','LaTeX','FontSize',24);
12   ylabel('$y$','Interpreter','LaTeX','FontSize',24);
13   % uncomment line below for EPS output
14   % print(h1,'-depsc','-r600','chap1Figure-MATLABStraightLinePlot.eps');
```

Appendix C
Chapter 2 VHDL, Simulink DSP Builder and SDC File

C.1 VHDL Generic Full Adder

Listing C.1 VHDL full adder with generics

```
1   library ieee;
2   use ieee.std_logic_1164.all;
3   use ieee.numeric_std.all;
4
5   entity genericNBitFullAdder is
6       generic (bitLength : integer := 0);
7       port (
8           xInBits,yInBits : in std_logic_vector(bitLength downto 0);
9           cIn : in std_logic;
10          cOut : out std_logic;
11          sumBits : out std_logic_vector(bitLength downto 0));
12  end genericNBitFullAdder;
13
14  architecture structuralRippleCarryAdder of genericNBitFullAdder is
15      component oneBitFullAdder is port (
16          xIn,yIn,cIn : in std_logic;
17          sum,cOut : out std_logic);
18      end component;
19      signal carryInternal : std_logic_vector(bitLength+1 downto 0);
20  begin
21      -- map cIn and cOut to carryInternal!
22      carryInternal(0) <= cIn;
23      cOut <= carryInternal(bitLength+1);
24      -- Generate statement for instantiating repeated structures
25      -- p. 799 (Appendix A) in your book (Fundamentals of Digital
26      -- Logic Design with VHDL)
27      generateAdders : -- generate label
28      for i IN 0 to bitLength generate
29          nFullAdders : oneBitFullAdder
30          port map (xInBits(i),yInBits(i),carryInternal(i),sumBits(i),
31          carryInternal(i+1));
32
33      end generate;
34  end structuralRippleCarryAdder;
```

© Springer International Publishing Switzerland 2015
B. Muthuswamy and S. Banerjee, *A Route to Chaos Using FPGAs*, Emergence, Complexity and Computation 16, DOI 10.1007/978-3-319-18105-9

Points to note from the VHDL description are:

1. In line 6 we have utilized the `generic` keyword to emphasize that, the size of the full adder can be resolved only at synthesis time.
2. Line 19 declares internal carry signals that will be utilized to interconnect the carry inputs and outputs of the one bit adders.
3. Lines 22 and 23 map the input carry to the least significant bit of the internal carry signal and the output carry to the most significant bit of the internal carry signal.
4. Lines 27–31 implement parameterization in VHDL via the for loop construct. Note that you need to make sure the VHDL syntax is correct, we have added carriage returns for code clarity.

C.2 VHDL Seven Segment Decoder

Listing C.2 VHDL behavioural seven segment decoder

```
 1  library ieee;
 2  use ieee.std_logic_1164.all;
 3  use ieee.numeric_std.all;
 4
 5  entity sevenSegmentDecoder is port (
 6      integerIn : in integer range 0 to 9;
 7      hexOut : out std_logic_vector(7 downto 0));
 8  end sevenSegmentDecoder;
 9
10  architecture behavioral of sevenSegmentDecoder is
11  begin
12      with integerIn select
13          hexOut <= X"40" when 0,
14                    X"79" when 1,
15                    X"24" when 2,
16                    X"30" when 3,
17                    X"19" when 4,
18                    X"12" when 5,
19                    X"02" when 6,
20                    X"78" when 7,
21                    X"00" when 8,
22                    X"10" when others;
23  end behavioral;
```

C.3 Top-Level for Generic Full Adder

Listing C.3 VHDL top level for generic adder

```vhdl
1   library ieee;
2   use ieee.std_logic_1164.all;
3   use ieee.numeric_std.all;
4
5   entity rippleCarryAdder is port (
6       SW : in std_logic_vector(9 downto 0);
7       HEX3,HEX2,HEX1,HEX0 : out std_logic_vector(6 downto 0);
8       -- LEDs are going to indicate carry out
9       LEDG : out std_logic_vector(7 downto 0));
10  end rippleCarryAdder;
11
12  architecture topLevel of rippleCarryAdder is
13
14      component genericNBitFullAdder is
15          generic (bitLength : integer := 0);
16          port (
17              xInBits,yInBits : in std_logic_vector(bitLength
                    downto 0);
18              cIn : in std_logic;
19              cOut : out std_logic;
20              sumBits : out std_logic_vector(bitLength downto 0));
21      end component;
22
23      component sevenSegmentDecoder is port (
24          integerIn : in integer range 0 to 9;
25          hexOut : out std_logic_vector(7 downto 0));
26      end component;
27
28      signal carry0 : std_logic;
29      signal sumInternal : std_logic_vector(3 downto 0);
30      signal sumInteger : integer;
31      signal unitsDigit,tensDigit : integer;
32      signal hex0Out,hex1Out : std_logic_vector(7 downto 0);
33
34  begin
35      carry0 <= '0';
36      oneBitFullAdder : genericNBitFullAdder generic map
37          (bitLength => 0)
38          port map ( xInBits => SW(0 downto 0),
39                     yInBits => SW(1 downto 1),
40                     cIn => carry0,
41                     cOut => LEDG(0),
42                     sumBits => LEDG(1 downto 1));
43      fourBitFullAdder : genericNBitFullAdder generic map
44          (bitLength => 3)
45          port map ( xInBits => SW(5 downto 2),
46                     yInBits => SW(9 downto 6),
```

```
47                          cIn => carry0,
48                          cOut => LEDG(2),
49                          sumBits => sumInternal(3 downto 0));
50
51      sumInteger <= to_integer(unsigned(sumInternal));
52      unitsDigit <= sumInteger rem 10;
53      tensDigit <= sumInteger / 10;
54
55      unitsDisplay : sevenSegmentDecoder port map (unitsDigit,
            hex0Out);
56      tensDisplay : sevenSegmentDecoder port map (tensDigit,
            hex1Out);
57      HEX0 <= hex0Out(6 downto 0);
58      HEX1 <= hex1Out(6 downto 0);
59
60  end topLevel;
```

1. Lines 36–47 show that we are going to realize two adders : an one bit full adder and a four bit full adder. The important VHDL syntax nuance is: since the genericNBitFullAdder module expects a std_logic_vector, we need to make sure our one bit inputs and outputs are of type std_logic_vector, not of type std_logic. Hence, instead of using SW(0) we use SW(0 down to 0).
2. Lines 49–51 convert the output from the four bit full adder into an unsigned integer and the digits from the integer are extracted for use with the sevenSegmentDecoder module. The online reference design video for the ALU realization in Sect. 2.3.2.2 has details on extracting digits from unsigned integers.

C.4 Seconds Counter with Single Pulse Generator

Listing C.4 VHDL seconds counter
```
1   library ieee;
2   use ieee.std_logic_1164.all;
3   use ieee.numeric_std.all;
4
5   entity secondsCounter is port (
6           areset,clockIn : in std_logic;
7           minuteCounterEnablePulse : out std_logic;
8           secondsOut : out integer range 0 to 59);
9   end entity;
10
11  architecture behavioralSecondsCounter of secondsCounter is
12
13      type state is (reset,generatePulse,stop);
14      signal currentState,nextState : state;
15
16      signal secondsCountRegister : integer range 0 to 49999999;
17      signal internalSecondsCount : integer range 0 to 59;
18      signal enablePulseGenerator : std_logic;
19  --
```

```vhdl
20  begin
21
22      secondsCountProcess : process(areset,clockIn)
23      begin
24          if areset='1' then
25              secondsCountRegister <= 0;
26              internalSecondsCount <= 0;
27              enablePulseGenerator <= '0';
28          else
29              if rising_edge(clockIn) then
30                  if secondsCountRegister >= 49999 then
31                      secondsCountRegister <= 0;
32                      if internalSecondsCount >= 59 then
33                          internalSecondsCount <= 0;
34                          enablePulseGenerator <= '1';
35                      else
36                          internalSecondsCount <=
                                  internalSecondsCount + 1;
37                          enablePulseGenerator <= '0';
38                      end if;
39                  else
40                      secondsCountRegister <= secondsCountRegister
                            +1;
41                  end if;
42
43              end if;
44          end if;
45      end process;
46
47      -- single pulse generator state machine
48      stateMemory : process(areset,clockIn)
49      begin
50          if areset='1' then
51              currentState <= reset;
52          else
53              if rising_edge(clockIn) then
54                  currentState <= nextState;
55              end if;
56          end if;
57      end process;
58      stateTransitionLogic : process(enablePulseGenerator,
            currentState)
59      begin
60          case currentState is
61              when reset =>
62                  if enablePulseGenerator='1' then
63                      nextState <= generatePulse;
64                  else
65                      nextState <= reset;
66                  end if;
67              when generatePulse =>
68                  nextState <= stop;
69              when others =>
```

```
70                      if enablePulseGenerator='0' then
71                          nextState <= reset;
72                      else
73                          nextState <= stop;
74                      end if;
75          end case;
76      end process;
77      -- output logic (could put this inside state transition
                logic process)
78      with currentState select
79          minuteCounterEnablePulse <= '1' when generatePulse,
80                                      '0' when others;
81      -- end single pulse generator FSM
82
83      --output seconds count register content for display
84      secondsOut <= internalSecondsCount;
85
86  end behavioralSecondsCounter;
```

The main ideas in the snippet that correspond to each of the blocks in Fig. 2.15 are:

1. Lines 13 and 14 illustrate how to use the type specification in VHDL to specify user-defined states. This will help the synthesizer (and simulator) infer a state machine from your design.
2. Lines 48–57 will infer the state memory block in Fig. 2.15.
3. Lines 58–76 will infer the next state logic block in Fig. 2.15.
4. Line 78 will infer the output logic block in Fig. 2.15.

C.5 Abstracting the FPGA Development Flow in Simulink

Simulink should be the tool of choice in realizing abstract mathematical concepts. For realizing chaotic system nonlinearities, we will utilize DSP builder blockset for Simulink from Altera. The reference for this section is Altera's DSP Builder Advanced Blockset reference manual that can be found on Altera's DSP Builder website [1].

1. The concept behind DSP builder is to create a synthesizable subsystem that incorporates our mathematical abstraction. To do so, we first access the **Altera DSP Builder Advanced Blockset**[2] library from Simulink, as shown in Fig. C.1.
2. We create a new Simulink model and then place the **Control** and **Signals** block from the **Base Blocks** library in Fig. C.1. Figures C.2 and C.3 show the configuration parameters for these blocks that will be used in a majority of the designs in this book.
3. We next set the solver configurations as shown in Fig. C.4. The completed top level is shown in Fig. C.5. We have created a subsystem at the top level. Within this subsystem, we will realize the nonlinear synthesizable subsystem.

---

[2] Altera recommends the use of Advanced Blockset instead of Standard Blockset for newer designs.

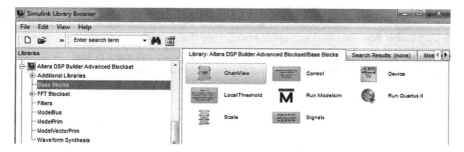

Fig. C.1 The Base blocks library in the DSP Builder Advanced Blockset

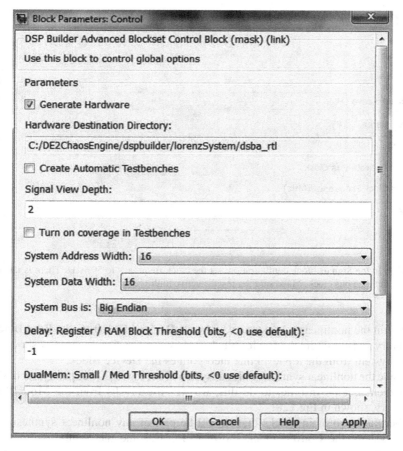

Fig. C.2 In the Control block, make sure generate hardware is checked and use an absolute path for the hardware destination directory. Turn off automatic test benches. Set both address and data width to 16-bits

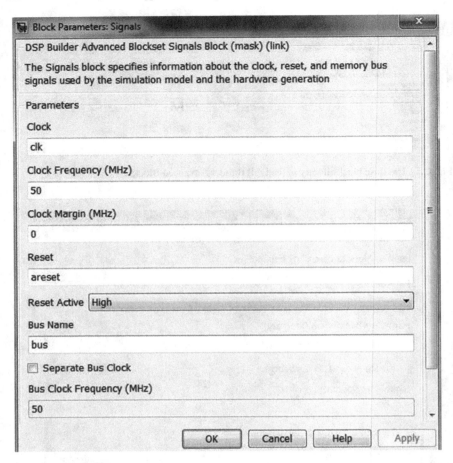

Fig. C.3 In the Signals block configuration, set the clock frequency to 50 MHz. There is no need to use a separate bus clock. Make sure the reset is active high

4. Within the nonlinear subsystem, we place a **Device** block from the **Base Blocks** library and configure the **Device** block as shown in Fig. C.6. Figure C.7 shows the subsystem from the top level that incorporates the **Device** Block.
5. Inside the nonlinear synthesizable subsystem from Fig. C.7, we specify our design mathematically. In order to do this, we will use blocks from the **ModelPrim** library, shown in Fig. C.8.
6. There are two main blocks that should be part of any nonlinear synthesizable subsystem: the Convert block shown in Fig. C.9 and the SynthesisInfo block shown in Fig. C.10.
7. We finally enable port data types as shown in Fig. C.11 as a visual debugging tool, in case of errors.

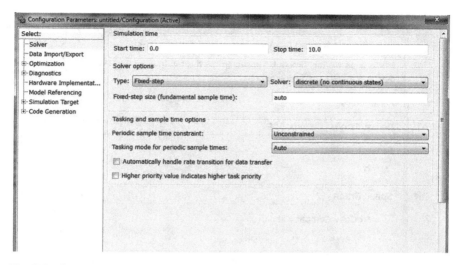

Fig. C.4 The solver should be configured as fixed-step and with no discrete states

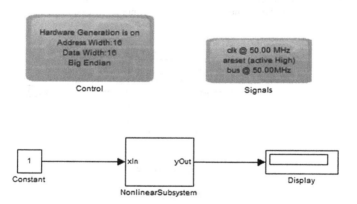

Fig. C.5 The top level of our design. According to DSP Builder syntax, the synthesizable portion of our design must be specified as a subsystem within the top level

Once we run our top level in Simulink, DSP builder should generate the appropriate hardware in the directory specified via the Control block (Fig. C.2). Make sure the constant input(s) from the Source library at the top level have single as the output data type. The default is double and will generate 64-bit bus widths.

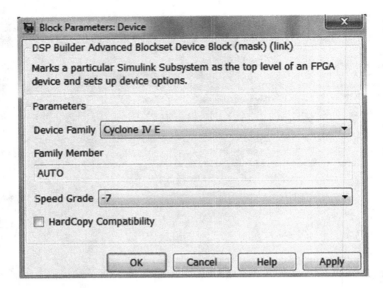

Fig. C.6 Device block configuration. We don't have to explicitly set the Cyclone IV E part number since we are only going to be synthesizing a subsystem, not a stand-alone Quartus project from DSP Builder

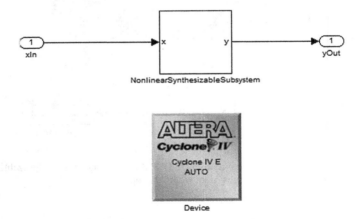

Fig. C.7 The Device block must be placed within a subsystem, not at the top level that has the Control and Signals blocks

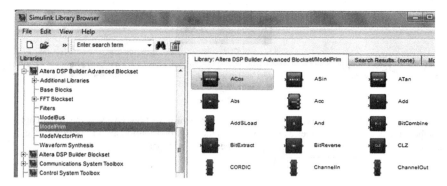

Fig. C.8 The ModelPrim library from the DSP Builder Advanced Blockset

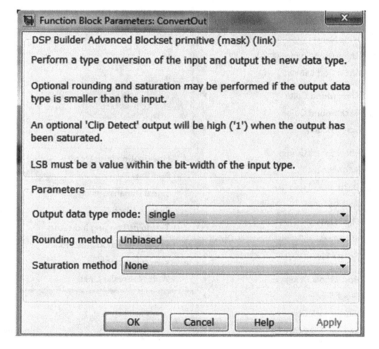

Fig. C.9 This block is used to convert inputs and outputs to single precision (32-bit) floating point

Fig. C.10 We need a SynthesisInfo block for controlling synthesis flow. By using the default option of Scheduled, we let DSP builder use a pipelining and delay distribution algorithm that create fast hardware implementations from an easily described untimed block diagram

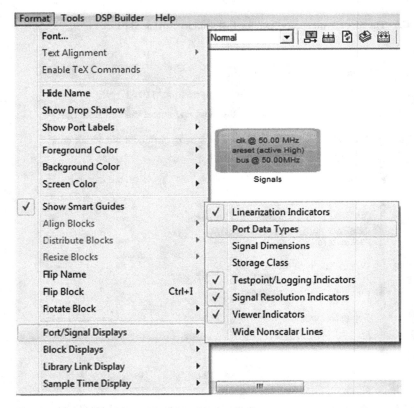

Fig. C.11 Enable port data types helps us debug the design more easily. In our experience, the most common error is incorrect data types. In the case of incorrect data types, functionally speaking, we have domain and range errors

C.6 SDC File for Twenty Four Hour Clock Design

Listing C.5 Synopsys Design Constraint File

```
 1  # STEP 1: ADD CLOCK CONSTRAINTS
 2  # the constraint below assigns a 20 ns clock to the input
      port
 3  # CLOCK_50. Note that we use the default name (CLOCK_50)
 4  # as the name of the clock. This is our board clock.
 5  create_clock -name CLOCK_50 -period 20.000
 6
 7  # There are no virtual I/O clocks since we are not driving
 8  # external devices. Moreover, because we have a SINGLE
 9  # clock for our entire design, virtual clocks are un-
      necessary.
10
11  # take care of PLL clocking constraints automatically.
12  # Note that this is an Altera SDC extension and this may not
13  # be available from all FPGA vendors. The create_base_clocks
14  # flag below generates create_clock constraints for PLL input
15  # clocks
16  derive_pll_clocks -create_base_clocks
17
18  # the command below automatically derives clock
      uncertainities
19  # such as setup, hold etc. Results only seen after place and
      route.
20  derive_clock_uncertainty
21
22  # Note that since we have a globally synchronous design
23  # (all flip-flops are being driven by the SAME PLL clock),
24  # we are DONE with clocking constraints!
25  # Once are done with clocking constraints, run TimeQuest
26  # to make sure clocks are constrainted.
27
28  # STEP 2: Add I/O constraints. We will add output constraints
29  # to the hex displays since these are actually registered
      outputs,
30  # with the global 50 MHz clock.
31  # The input and output delays really don't matter for this
      example.
32  # Of course, if our design is driving timing critical
      external logic,
33  # then the board clock skew etc. need to be accurate.
34  set_output_delay -clock { CLOCK_50 } -max 5 [get_ports {HEX
      *}]
35  set_output_delay -clock { CLOCK_50 } -min 10 [get_ports {HEX
      *}]
36
37  # STEP 3: Asynchronous path(s). Since the external KEY(0)
      input
38  # is truly asynchronous, we have internally synchronized the
      reset.
```

```
39  # Hence, we can safely apply a false path constraint.
40  set_false_path -from [get_ports KEY*]
```

Reference

1. Altera Corporation (2013) DSP Builder, Available via DIALOG. http://www.altera.com/produc
ts/software/products/dsp/dsp-builder.html. Accessed 7 May 2013

Appendix D
Chapter 3 VHDL, MATLAB Code and ModelSim Scripts

D.1 VHDL Specification of the Lorenz System

Listing D.1 VHDL specification of the Lorenz system. We have split comments over multiple lines so be careful when copying and pasting the HDL! Note that the specification is incomplete, we use it to understand the ideas behind Euler's method

```
1   entity lorenzSystem is port (
2       resetn, clockIn : in std_logic;
3       xOut,yOut,zOut : out std_logic_vector(15 downto 0));
4   end lorenzSystem;
5
6   architecture behavioral of lorenzSystem is
7
8   signal reset : std_logic;
9   ...
10  -- state variables
11  signal x1,x2,x3,x1New,x2New,x3New : std_logic_vector (31 downto 0)
        ;
12  ...
13  begin
14      reset <= not resetn;
15      -- Euler's method
16      ...
17      -- state memory
18      process(clockIn, resetn)
19      begin
20              -- constants (place outside reset and clock to avoid
                    latches)
21          ...
22          if resetn = '0' then
23              -- initial state
24              -- the constants below are in single-precision 32-bit
                    floating point format
25              -- You can use an online floating point converter such
                    as :
26              -- http://babbage.cs.qc.cuny.edu/IEEE-754.old/Decimal.
                    html
```

© Springer International Publishing Switzerland 2015
B. Muthuswamy and S. Banerjee, *A Route to Chaos Using FPGAs*, Emergence,
Complexity and Computation 16, DOI 10.1007/978-3-319-18105-9

```
27              -- to determine the appropriate 32-bit hexadecimal
                   values for the corresponding reals.
28              x1 <= X"41200000";-- 10
29              x2 <= X"41A00000";-- 20
30              x3 <= X"41F00000";-- 30
31              ...
32          else
33              if rising_edge(clockIn) then
34                  ... -- appropriate state update computations go
                       here (will be discussed later)
35                  x1 <= x1New;
36                  x2 <= x2New;
37                  x3 <= x3New;
38              end if;
39          end if;
40      end process;
41      ...
42      ...
43  end behavioral;
```

D.2 VHDL Specification of the Lorenz System with Mathematical Labelling of Signals

Listing D.2 VHDL specification of the Lorenz system with consistent mathematical labels for the various signals. This specification is also incomplete

```
1  entity lorenzSystem is port (
2       resetn, clockIn : in std_logic;
3       xOut,yOut,zOut : out std_logic_vector(15 downto 0));
4  end lorenzSystem;
5
6  architecture behavioral of lorenzSystem is
7
8  signal reset : std_logic;
9  ...
10 -- state variables
11 signal x,y,z,xNew,yNew,zNew : std_logic_vector(31 downto 0);
12 ...
13 begin
14     reset <= not resetn;
15     -- Euler's method
16     ...
17     -- state memory
18     process(clockIn, resetn)
19     begin
20              -- constants (place outside reset and clock to
                   avoid latches)
21              ...
22              if resetn = '0' then
23                  -- initial state
```

```
24          -- the constants below are in single-precision
               32-bit floating point format
25          -- You can use an online floating point converter
               such as :
26          -- http://babbage.cs.qc.cuny.edu/IEEE-754.old/
               Decimal.html
27          -- to determine the appropriate 32-bit
               hexadecimal values for the corresponding
               reals.
28          x <= X"41200000";-- 10
29          y <= X"41A00000";-- 20
30          z <= X"41F00000";-- 30
31          ...
32        else
33          if rising_edge(clockIn) then
34              ... -- appropriate state update computations
                   go here (will be discussed later)
35              x <= xNew;
36              y <= yNew;
37              z <= zNew;
38          end if;
39        end if;
40      end process;
41      ...
42      ...
43  end behavioral;
```

D.3 MATLAB Code for Plotting Lorenz System Trajectory Obtained from Euler's Method

Listing D.3 MATLAB code for plotting output from Simulink
```
1  >> x=xSimOut.signals.values; % extract x data from structure
2  >> t=linspace(0,1,1001);
3  >> plot(t,x')
4  >> xlabel('t');
5  >> ylabel('x(t)');
6  >> title('x(t) plot for Lorenz System');
```

D.4 Complete VHDL Specification of the Lorenz System

Listing D.4 Complete VHDL specification of the Lorenz system, incorporating both DSP builder synthesized subsystem and Euler's method

```
 1  library ieee;
 2  use ieee.std_logic_1164.all;
 3  use ieee.std_logic_signed.all;
 4
 5  entity lorenzSystem is port (
 6        resetn, clockIn : in std_logic;
 7        xOut,yOut,zOut : out std_logic_vector(15 downto 0));
 8  end lorenzSystem;
 9
10  architecture behavioral of lorenzSystem is
11
12  signal reset : std_logic;
13  -- constants
14  signal dt,output1Over10Scale,output1Over20Scale,
           output1Over30Scale : std_logic_vector(31 downto 0);
15  -- state variables
16  signal x,y,z,xNew,yNew,zNew,xScaled,yScaled,zScaled,xFixed,
           yFixed,zFixed : std_logic_vector(31 downto 0);
17  -- prescalar
18  signal count : integer range 0 to 64;
19
20  -- DSP builder top level.
21  -- Note: Open lorenzSystem_NonlinearSubsystem.vhd and create
           component.
22  -- NOTE : SAME STEPS FOR OTHER CHAOTIC SYSTEMS!
23  component lorenzSystem_NonlinearSubsystem is
24    port (
25        In_dt : in std_logic_vector(31 downto 0);
26        In_x : in std_logic_vector(31 downto 0);
27        In_y : in std_logic_vector(31 downto 0);
28        In_z : in std_logic_vector(31 downto 0);
29        Out_x : out std_logic_vector(31 downto 0);
30        Out_y : out std_logic_vector(31 downto 0);
31        Out_z : out std_logic_vector(31 downto 0);
32        clk : in std_logic;
33        areset : in std_logic;
34        h_areset : in std_logic
35        );
36  end component;
37  -- END DSP builder top level.
38
39  -- latency : 5 clock cycles (scale for DAC range)
40  component floatingPointMultiplyDedicated IS
41    PORT
42
43        aclr : IN STD_LOGIC ;
44        clock : IN STD_LOGIC ;
45        dataa : IN STD_LOGIC_VECTOR (31 DOWNTO 0);
```

```vhdl
46          datab : IN STD_LOGIC_VECTOR (31 DOWNTO 0);
47          result : OUT STD_LOGIC_VECTOR (31 DOWNTO 0)
48      );
49  END component;
50
51
52  -- latency : 6 clock cycles
53  component floatingPointToFixedLorenz IS
54      PORT
55      (
56          aclr : IN STD_LOGIC ;
57          clock : IN STD_LOGIC ;
58          dataa : IN STD_LOGIC_VECTOR (31 DOWNTO 0);
59          result : OUT STD_LOGIC_VECTOR (31 DOWNTO 0)
60      );
61  END component;
62
63  component floatingPointToFixedZLorenz IS
64      PORT
65      (
66          aclr : IN STD_LOGIC ;
67          clock : IN STD_LOGIC ;
68          dataa : IN STD_LOGIC_VECTOR (31 DOWNTO 0);
69          result : OUT STD_LOGIC_VECTOR (31 DOWNTO 0)
70      );
71  END component;
72
73
74  begin
75      reset <= not resetn;
76      -- Euler's method
77      -- We first synchronously update state variables at 781.250
                KHz (64 counts of 50 MHz clock)
78      -- Since dt = 1/1024, time scale is actually (780.250e3/1024)
                = 762 Hz (approximately)
79
80      -- state memory
81      process(clockIn, resetn)
82      begin
83              -- constants (place outside reset and clock to avoid
                    latches)
84              dt <= X"3A800000"; -- 1/1024
85              output1Over10Scale <= X"3DCCCCCC"; --0.1
86              output1Over20Scale <= X"3D4CCCCC"; -- 0.05
87              output1Over30Scale <= X"3D888888"; -- 0.06666 approx.
88          if resetn = '0' then
89              -- initial state
90              x <= X"41200000"; -- 10
91              y <= X"41A00000"; -- 20
92              z <= X"41F00000"; -- 30
93              count <= 0;
94          else
95              if rising_edge(clockIn) then
```

```vhdl
 96               if count = 64 then
 97                   count <= 0;
 98               else
 99                   count <= count + 1;
100               end if;
101
102               if count = 63 then
103                   x <= xNew;
104                   y <= yNew;
105                   z <= zNew;
106               end if;
107           end if;
108       end if;
109   end process;
110
111   staticNonlinearitiesAndDeltaT :
          lorenzSystem_NonlinearSubsystem port map (
112        In_dt => dt,
113        In_x => x,
114        In_y => y,
115        In_z => z,
116        Out_x => xNew,
117        Out_y => yNew,
118        Out_z => zNew,
119        clk => clockIn,
120        areset => reset,
121        h_areset => reset);
122   -- END Euler's method
123
124   -- scale outputs
125   scaleX : floatingPointMultiplyDedicated port map (
126        aclr => reset,
127        clock => clockIn,
128        dataa => x,
129        datab => output1Over10Scale,
130        result => xScaled);
131   scaleY : floatingPointMultiplyDedicated port map (
132        aclr => reset,
133        clock => clockIn,
134        dataa => y,
135        datab => output1Over20Scale,
136        result => yScaled);
137   scaleZ : floatingPointMultiplyDedicated port map (
138        aclr => reset,
139        clock => clockIn,
140        dataa => z,
141        datab => output1Over30Scale,
142        result => zScaled);
143
144   --state outputs : convert scaled floating point x,y variables
             to 2.30 fixed point for DAC
145   xOutFinal : floatingPointToFixedLorenz port map (
146        aclr => reset,
```

```
147          clock => clockIn,
148          dataa => xScaled,
149          result => xFixed);
150     yOutFinal : floatingPointToFixedLorenz port map (
151          aclr => reset,
152          clock => clockIn,
153          dataa => yScaled,
154          result => yFixed);
155     -- convert scaled z variable to 3.29 fixed point since z(t)
                for Lorenz requires larger resolution for the
156     -- magnitude component (this should be evident from the
                MATLAB and/or ModelSim simulation.
157     zOutFinal : floatingPointToFixedZLorenz port map (
158          aclr => reset,
159          clock => clockIn,
160          dataa => zScaled,
161          result => zFixed);
162
163          xOut <= xFixed(31 downto 16);
164          yOut <= yFixed(31 downto 16);
165          zOut <= zFixed(31 downto 16);
166
167   end behavioral;
```

D.5   Complete VHDL Specification of the Highly Complex Attractor System

Listing D.5 Highly Complex Attractor System in VHDL

```
1    --04/10/13
2    --Single precision (32-bit) floating point realization
3
4    library ieee;
5    use ieee.std_logic_1164.all;
6    use ieee.std_logic_signed.all;
7
8    entity highlyComplexAttractorSystem is port (
9         resetn, clockIn : in std_logic;
10        xOut,yOut,zOut : out std_logic_vector(15 downto 0));
11   end highlyComplexAttractorSystem;
12
13   architecture behavioral of highlyComplexAttractorSystem is
14
15   signal reset : std_logic;
16   -- constants
17   signal dt,outputHalfScale,output1Over8Scale :
             std_logic_vector(31 downto 0);
18   -- state variables
19   signal x,y,z,xNew,yNew,zNew,xScaled,yScaled,zScaled,xFixed,
             yFixed,zFixed : std_logic_vector(31 downto 0);
```

```vhdl
20  -- prescalar
21  signal count : integer range 0 to 64;
22
23  -- DSP builder top level. Steps to add:
24  -- 1. source ./dspba_rtl/highlyComplexAttractor/
        NonlinearSubsystem/NonlinearSubsystem.add.tcl via TCL
        window (View -> Utility Windows -> TCL Console)
25  -- 2. source ./dspba_rtl/highlyComplexAttractor/
        NonlinearSubsystem/
        highlyComplexAttractor_NonlinearSubsystem_fpc.add.tcl
        via TCL window (View -> Utility Windows -> TCL Console)
26  -- 2  Open highlyComplexAttractor_NonlinearSubsystem.vhd
        from the path above and create component.
27  -- NOTE : SAME STEPS FOR OTHER CHAOTIC SYSTEMS!
28  component highlyComplexAttractor_NonlinearSubsystem is
29    port (
30        In_dt : in std_logic_vector(31 downto 0);
31        In_x : in std_logic_vector(31 downto 0);
32        In_y : in std_logic_vector(31 downto 0);
33        In_z : in std_logic_vector(31 downto 0);
34        Out_x : out std_logic_vector(31 downto 0);
35        Out_y : out std_logic_vector(31 downto 0);
36        Out_z : out std_logic_vector(31 downto 0);
37        clk : in std_logic;
38        areset : in std_logic;
39        h_areset : in std_logic);
40  end component;
41  -- END DSP builder top level.
42
43  -- latency : 5 clock cycles (scale for DAC range)
44  component floatingPointMultiplyDedicated IS
45    PORT
46      (
47        aclr : IN STD_LOGIC ;
48        clock : IN STD_LOGIC ;
49        dataa : IN STD_LOGIC_VECTOR (31 DOWNTO 0);
50        datab : IN STD_LOGIC_VECTOR (31 DOWNTO 0);
51        result : OUT STD_LOGIC_VECTOR (31 DOWNTO 0)
52      );
53  END component;
54
55
56  -- latency : 6 clock cycles
57  component floatingPointToFixed IS
58    PORT
59      (
60        aclr : IN STD_LOGIC ;
61        clock : IN STD_LOGIC ;
62        dataa : IN STD_LOGIC_VECTOR (31 DOWNTO 0);
63        result : OUT STD_LOGIC_VECTOR (31 DOWNTO 0)
64      );
65  END component;
66
```

```
67  begin
68     reset <= not resetn;
69     -- Euler's method
70     -- We first synchronously update state variables at
          781.250 KHz (64 counts of 50 MHz clock)
71     -- Since dt = 1/1024, time scale is actually (780.250e3
          /1024) = 762 Hz (approximately)
72
73     -- state memory
74     process(clockIn, resetn)
75     begin
76              -- constants (place outside reset and clock to
                   avoid latches)
77             dt <= X"3A800000"; -- 1/1024
78             outputHalfScale <= X"3F000000"; --0.5
79             output1Over8Scale <= X"3E000000"; -- 0.125
80          if resetn = '0' then
81             -- initial state
82             x <= X"3F800000"; -- 1
83             y <= X"3F800000";
84             z <= X"3F800000";
85             count <= 0;
86          else
87             if rising_edge(clockIn) then
88                if count = 64 then
89                   count <= 0;
90                else
91                   count <= count + 1;
92                end if;
93
94                if count = 63 then
95                   x <= xNew;
96                   y <= yNew;
97                   z <= zNew;
98                end if;
99             end if;
100         end if;
101     end process;
102
103     staticNonlinearities :
           highlyComplexAttractor_NonlinearSubsystem port map (
104        In_dt => dt,
105        In_x => x,
106        In_y => y,
107        In_z => z,
108        Out_x => xNew,
109        Out_y => yNew,
110        Out_z => zNew,
111        clk => clockIn,
112        areset => reset,
113        h_areset => reset);
114     -- END Euler's method
115
```

```
116    -- scale outputs
117    scaleX : floatingPointMultiplyDedicated port map (
118        aclr => reset,
119        clock => clockIn,
120        dataa => x,
121        datab => outputHalfScale,
122        result => xScaled);
123    scaleY : floatingPointMultiplyDedicated port map (
124        aclr => reset,
125        clock => clockIn,
126        dataa => y,
127        datab => outputHalfScale,
128        result => yScaled);
129    scaleZ : floatingPointMultiplyDedicated port map (
130        aclr => reset,
131        clock => clockIn,
132        dataa => z,
133        datab => output1Over8Scale,
134        result => zScaled);
135
136    --state outputs : convert scaled floating point variables
               to 5.27 fixed point format DAC (no latency)
137    xOutFinal : floatingPointToFixed port map (
138        aclr => reset,
139        clock => clockIn,
140        dataa => xScaled,
141        result => xFixed);
142    yOutFinal : floatingPointToFixed port map (
143        aclr => reset,
144        clock => clockIn,
145        dataa => yScaled,
146        result => yFixed);
147    zOutFinal : floatingPointToFixed port map (
148        aclr => reset,
149        clock => clockIn,
150        dataa => zScaled,
151        result => zFixed);
152
153        xOut <= xFixed(31 downto 16);
154        yOut <= yFixed(31 downto 16);
155        zOut <= zFixed(31 downto 16);
156
157    end behavioral;
```

D.6 VHDL Testbench for Chen System

Listing D.6 A test bench. Note that test benches are not synthesizable

```
1    -- testbench for Chen system
2
3    library ieee;
```

```vhdl
 4  use ieee.std_logic_1164.all;
 5  use ieee.numeric_std.all;
 6
 7  entity chenSystemtb is
 8  end chenSystemtb;
 9
10  architecture testbench of chenSystemtb is
11      signal clock,reset,resetn,trigger,increment,pulseOut :
              std_logic := '0';
12      signal xOut,yOut,zOut : std_logic_vector(15 downto 0);
13
14      component chenSystem is port (
15          resetn, clockIn : in std_logic;
16          xOut,yOut,zOut : out std_logic_vector(15 downto 0));
17      end component;
18
19  begin
20      chenSystemInstance : chenSystem port map (
21          resetn => resetn,
22          clockIn => clock,
23          xOut => xOut,
24          yOut => yOut,
25          zOut => zOut);
26
27      clock50MHzProcess : process
28      begin
29          clock <= not clock;
30          wait for 10 ns;
31      end process clock50MHzProcess;
32
33      stimulus : process
34      begin
35          resetn <= '0';
36          wait for 55 ns;
37          resetn <= '1'; -- unreset after 55 ns
38          wait; -- prevent process from being executed again
39      end process stimulus;
40  end testbench;
```

The salient features of the test bench are:

1. The entity statement has no input and/or output ports. This makes sense since the test bench is a virtual environment that cannot be synthesized.
2. We have to generate a 50 MHz clock. This is done using the process statement shown.
3. We then provide stimulus inputs to our module under test. Notice that since VHDL processes execute concurrently, our stimulus process has a wait as the last statement to prevent repeated process execution.

D.7 ModelSim Script File for Chen System

Listing D.7 Script file for Chen system that lists ModelSim commands for performing functional simulation

```
1   # do vlib work only once!
2   # vlib work
3
4   vmap work work
5
6   # compile DSP builder advanced blockset source
7   vcom D:/altera/12.0/quartus/dspba/Libraries/vhdl/fpc/math_package.vhd
8   vcom D:/altera/12.0/quartus/dspba/Libraries/vhdl/fpc/math_implementation.vhd
9   vcom D:/altera/12.0/quartus/dspba/Libraries/vhdl/fpc/hcc_package.vhd
10  vcom D:/altera/12.0/quartus/dspba/Libraries/vhdl/fpc/hcc_implementation.vhd
11  vcom D:/altera/12.0/quartus/dspba/Libraries/vhdl/fpc/fpc_library_package.vhd
12  vcom D:/altera/12.0/quartus/dspba/Libraries/vhdl/fpc/fpc_library.vhd
13
14  # compile DSP builder advanced blockset nonlinearities
15  vcom ../dspBuilder/chenSystem/dspba_rtl/chenSystem/NonlinearSubsystem/
        chenSystem_NonlinearSubsystem.vhd
16  vcom ../dspBuilder/chenSystem/dspba_rtl/chenSystem/NonlinearSubsystem/
        NonlinearSynthesizableSubsystem/
        chenSystem_NonlinearSubsystem_NonlinearSynthesizableSubsystem.vhd
17
18  # compile source
19  vcom ./floatingPointMultiplyDedicated.vhd
20  vcom ../floatingPointToFixed.vhd
21  vcom ../chenSystem.vhd
22
23
24  vcom chenSystemtb.vhd
25
26  vsim chenSystemtb
27  # configure wave window to have a white background color
28  # http://www.utdallas.edu/~zhoud/EE%203120/Xilinx_tutorial_Spartan3_home_PC.
        pdf and ModelSim Reference Manual - configure command
29  configure wave -background white -foreground red -textcolor blue -timecolor
        blue -vectorcolor red -wavebackground white -cursorcolor black
30  add wave -divider "Clock and Reset"
31  add wave clock
32  add wave resetn
33
34
35  add wave -divider "Outputs from Chen System model"
36  # obtained switch information below by using analog (automatic) formatting in
        ModelSim
37  add wave -format analog-step -min -13380 -max 11972 -height 74 xOut
38  add wave -format analog-step -min -26982 -max 24076 -height 74 yOut
39  add wave -format analog-step -min -11033 -max 20597 -height 74 zOut
40
41  add wave -divider "Chen System Module Internal Signals"
42  add wave -label dt -hex sim:/chenSystemtb/chenSystemInstance/dt
43  add wave -label count -hex sim:/chenSystemtb/chenSystemInstance/count
44  add wave -label xPlus_f1 -hex sim:/chenSystemtb/chenSystemInstance/xNew
45  add wave -label yPlus_f2 -hex sim:/chenSystemtb/chenSystemInstance/yNew
46  add wave -label zPlus_f3 -hex sim:/chenSystemtb/chenSystemInstance/zNew
47  add wave -label x -hex sim:/chenSystemtb/chenSystemInstance/x
48  add wave -label y -hex sim:/chenSystemtb/chenSystemInstance/y
49  add wave -label z -hex sim:/chenSystemtb/chenSystemInstance/z
50  add wave -label xFixed -hex sim:/chenSystemtb/chenSystemInstance/xFixed
```

```
51   add wave -label yFixed -hex sim:/chenSystemtb/chenSystemInstance/yFixed
52   add wave -label zFixed -hex sim:/chenSystemtb/chenSystemInstance/zFixed
53
54   # run 1ms
```

Appendix E
Chapter 4 MATLAB Code, VHDL and ModelSim Scripts

E.1 Rössler System Specification in MATLAB

Listing E.1 MATLAB code for the Rössler system

```matlab
1  function [t,y] = rossler(tspan,reltol,abstol,x0,alpha,beta,
       gamma)
2
3  % Simulates the Rossler system:
4  % x'=-y-z
5  % y'=x+alpha*y
6  % z'=beta+z*(x-gamma)
7  % Function uses ode45. The arguments to be passed into the
       function
8  % are tspan, reltol,abstol,x0,alpha,beta,gamma. For classic
       Rossler
9  % attractor, try:
10 % [t,rosslerOut]=rossler([0:0.01:100],1e-5,1e-5,[14.5 0
       0.1],0.1,0.1,14);
11
12 options = odeset('RelTol',reltol,'AbsTol',abstol);
13 [t,y] = ode45(@rosslerFunction,tspan,x0,options);
14
15
16    function dy = rosslerFunction(t,y)
17        dy = zeros(3,1); % a column vector
18        dy(1) = -y(2)-y(3);
19        dy(2) = y(1) + alpha*y(2);
20        dy(3) = beta+y(3)*(y(1)-gamma);
21    end
22 end
```

© Springer International Publishing Switzerland 2015
B. Muthuswamy and S. Banerjee, *A Route to Chaos Using FPGAs*, Emergence,
Complexity and Computation 16, DOI 10.1007/978-3-319-18105-9

E.2 Period-1 Limit Cycle for the Rössler System

Listing E.2 MATLAB code for obtaining a limit cycle (period 1) the Rössler system

```
 1   % type one line at a time or use a script file after defining the
         rossler function
 2   h1=figure;
 3   [t,y1]=rossler([0:0.01:100],1e-5,1e-5,[14.5 0 0.1],0.1,0.1,4);
 4   plot(y1([5000:10001],1),y1([5000:10001],2))
 5   a = get(gca,'XTickLabel');
 6   set(gca,'XTickLabel',a,'fontsize',18)
 7   xlabel('$y$','Interpreter','Latex','Fontsize',32)
 8   ylabel('$x$','Interpreter','Latex','Fontsize',32)
 9   % enable line below for EPS output
10   % print(h1,'-depsc','-r600','chap4Figure-
         RosslerSystemPeriodOneLimitCycle.eps');
```

E.3 MATLAB Code for Period-Doubling Route to Chaos

Listing E.3 MATLAB code for obtaining period-doubling bifurcation in the Rössler system

```
 1   % rossler period-doubling script
 2   % make sure rossler.m is in the same folder
 3   [t,rosslerPeriod3]=rossler([0:0.01:10000],1e-5,1e-5,[14.5 0
         0.1],0.1,0.1,12);
 4   [t,rosslerPeriod6]=rossler([0:0.01:10000],1e-5,1e-5,[14.5 0
         0.1],0.1,0.1,12.6);
 5   [t,rosslerPeriodHigh]=rossler([0:0.01:10000],1e-5,1e-5,[14.5
         0 0.1],0.1,0.1,13.3);
 6   [t,rosslerChaos]=rossler([0:0.01:10000],1e-5,1e-5,[14.5 0
         0.1],0.1,0.1,14);
 7
 8   h1 = figure;
 9   plot(rosslerPeriod3([75000:100001],1),rosslerPeriod3
         ([75000:100001],2))
10   a = get(gca,'XTickLabel');
11   set(gca,'XTickLabel',a,'fontsize',18)
12   xlabel('$x$','Interpreter','Latex','Fontsize',32)
13   ylabel('$y$','Interpreter','Latex','Fontsize',32)
14   title('Period=3','Interpreter','Latex','Fontsize',32)
15   % enable line below for EPS output
16   % print(h1,'-depsc','-r600','chap4Figure-rosslerP3.eps');
17
18   h2=figure;
19   plot(rosslerPeriod6([75000:100001],1),rosslerPeriod6
         ([75000:100001],2))
20   a = get(gca,'XTickLabel');
21   set(gca,'XTickLabel',a,'fontsize',18)
22   xlabel('$x$','Interpreter','Latex','Fontsize',32)
23   ylabel('$y$','Interpreter','Latex','Fontsize',32)
24   title('Period=6','Interpreter','Latex','Fontsize',32)
```

```
25   % enable line below for EPS output
26   % print(h2,'-depsc','-r600','chap4Figure-rosslerP6.eps');
27
28   h3=figure;
29   plot(rosslerPeriodHigh([75000:100001],1),rosslerPeriodHigh
         ([75000:100001],2))
30   a = get(gca,'XTickLabel');
31   set(gca,'XTickLabel',a,'fontsize',18)
32   xlabel('$x$','Interpreter','Latex','Fontsize',32)
33   ylabel('$y$','Interpreter','Latex','Fontsize',32)
34   title('Higher Period','Interpreter','Latex','Fontsize',32)
35   % enable line below for EPS output
36   % print(h3,'-depsc','-r600','chap4Figure-rosslerHighPeriod.
         eps');
37
38   h4=figure;
39   plot(rosslerChaos([75000:100001],1),rosslerChaos
         ([75000:100001],2))
40   a = get(gca,'XTickLabel');
41   set(gca,'XTickLabel',a,'fontsize',18)
42   xlabel('$x$','Interpreter','Latex','Fontsize',32)
43   ylabel('$y$','Interpreter','Latex','Fontsize',32)
44   title('Chaos','Interpreter','Latex','Fontsize',32)
45   % enable line below for EPS output
46   % print(h4,'-depsc','-r600','chap4Figure-rosslerChaos.eps');
```

E.4 MATLAB Code for Chua Oscillator

Listing E.4 MATLAB code for simulating Chua oscillator

```
1    function [t,y] = chuaOscillator(tspan,reltol,abstol,x0,alpha,
         beta,gamma,a,c)
2
3    % Simulates Chua's oscillator with a smooth nonlinearity
4    % from Ambelang's EE4060 project report (Spring 2011
         Nonlinear Dynamics
5    % Course at MSOE)
6    % x'=alpha*(y-g(x))
7    % y'=x-y+z
8    % z'=-beta*y-gamma*z
9    % Function uses ode45. The arguments to be passed into the
         function
10   % are tspan, reltol,abstol,x0,alpha,beta,,gamma,a and c. a
         and c are
11   % parameters for the nonlinear function:
12   % f(x)=-a*x+0.5(a+b)(|x+1|-|x-1|)
13   % For a double-scroll chaotic attractor, try:
14   % [t,doubleScroll]=chuaOscillator([0:0.1:1000],1e-5,1e-5,[0.1
         0 0.1],10,16,0,1,-0.143);
15   % For period-adding route,
```

```
16  % [t,period3]=chuaOscillator([0:0.1:1000],1e-5,1e-5,[0.1 0
        0.1],3.708,3.499,0.076,1,-0.276);
17  % [t,chaosAfterPeriod3]=chuaOscillator([0:0.1:1000],1e-5,1e
        -5,[0.1 0 0.1],3.708,3.549,0.076,1,-0.276);
18  % [t,period4]=chuaOscillator([0:0.1:1000],1e-5,1e-5,[0.1 0
        0.1],3.708,3.574,0.076,1,-0.276);
19  % [t,chaosAfterPeriod4]=chuaOscillator([0:0.1:1000],1e-5,1e
        -5,[0.1 0 0.1],3.708,3.6,0.076,1,-0.276);
20
21  options = odeset('RelTol',reltol,'AbsTol',abstol);
22  [t,y] = ode45(@chuaOscillatorFunction,tspan,x0,options);
23
24
25    function dy = chuaOscillatorFunction(t,y)
26        dy = zeros(3,1); % a column vector
27        dy(1) = alpha*(y(2)-g(y(1),a,c));
28        dy(2) = y(1)-y(2)+y(3);
29        dy(3) = -beta*y(2)-gamma*y(3);
30    end
31
32    function y=g(x,a,c)
33        y=a*x^3+c*x;
34    end
35  end
```

E.5 MATLAB Code for Period-Adding Route to Chaos

Listing E.5 MATLAB code for obtaining the period-adding route to chaos in Fig. 4.3
```
1  % period-adding, using Chua's oscillator
2  % Make sure chuaOscillator.m is in the same folder
3  [t,period3]=chuaOscillator([0:0.1:1000],1e-5,1e-5,[0.1 0
       0.1],3.708,3.499,0.076,1,-0.276);
4  [t,chaosAfterPeriod3]=chuaOscillator([0:0.1:1000],1e-5,1e-5,[0.1
       0 0.1],3.708,3.549,0.076,1,-0.276);
5  [t,period4]=chuaOscillator([0:0.1:1000],1e-5,1e-5,[0.1 0
       0.1],3.708,3.574,0.076,1,-0.276);
6  [t,chaosAfterPeriod4]=chuaOscillator([0:0.1:1000],1e-5,1e-5,[0.1
       0 0.1],3.708,3.6,0.076,1,-0.276);
7
8  h1 = figure;
9  plot(period3([5000:10001],1),period3([5000:10001],2))
10 a = get(gca,'XTickLabel');
11 set(gca,'XTickLabel',a,'fontsize',18)
12 xlabel('$y$','Interpreter','Latex','Fontsize',32)
13 ylabel('$x$','Interpreter','Latex','Fontsize',32)
14 title('3:3 Limit Cycle','Fontsize',32)
15 % enable line below for EPS output
16 % print(h1,'-depsc','-r600','chap4Figure-chuaOscillatorP3.eps');
17
18 h2 = figure;
```

```
19  plot(chaosAfterPeriod3([5000:10001],1),chaosAfterPeriod3
        ([5000:10001],2))
20  a = get(gca,'XTickLabel');
21  set(gca,'XTickLabel',a,'fontsize',18)
22  xlabel('$y$','Interpreter','Latex','Fontsize',32)
23  ylabel('$x$','Interpreter','Latex','Fontsize',32)
24  title('Chaos after 3:3 Limit Cycle ','Fontsize',32)
25  % enable line below for EPS output
26  % print(h2,'-depsc','-r600','chap4Figure-
        chuaOscillatorChaosAfterP3.eps');
27
28  h3 = figure;
29  plot(period4([5000:10001],1),period4([5000:10001],2))
30  a = get(gca,'XTickLabel');
31  set(gca,'XTickLabel',a,'fontsize',18)
32  xlabel('$y$','Interpreter','Latex','Fontsize',32)
33  ylabel('$x$','Interpreter','Latex','Fontsize',32)
34  title('4:4 Limit Cycle ','Fontsize',32)
35  % enable line below for EPS output
36  % print(h3,'-depsc','-r600','chap4Figure-chuaOscillatorP4.eps');
37
38  h4 = figure;
39  plot(chaosAfterPeriod4([5000:10001],1),chaosAfterPeriod4
        ([5000:10001],2))
40  a = get(gca,'XTickLabel');
41  set(gca,'XTickLabel',a,'fontsize',18)
42  xlabel('$y$','Interpreter','Latex','Fontsize',32)
43  ylabel('$x$','Interpreter','Latex','Fontsize',32)
44  title('Chaos after 4:4 Limit Cycle ','Fontsize',32)
45  % enable line below for EPS output
46  % print(h4,'-depsc','-r600','chap4Figure-
        chuaOscillatorChaosAfterP4.eps');
```

E.6 MATLAB Code for Torus-Breakdown System

Listing E.6 MATLAB code implementing Eqs. (4.8)–(4.10)

```
1   function [t,y] = torusBreakdown(tspan,reltol,abstol,x0,alpha,beta
        ,a,b)
2
3   % Simulates the torus breakdown system from Matsumoto et. al.:
4   % x'=-alpha*f(y-x)
5   % y'=-f(y-x)-z
6   % z'=beta*y
7   % Function uses ode45. The arguments to be passed into the
        function
8   % are tspan, reltol,abstol,x0,alpha,beta,a and b. a and b are
        parameters
9   % for the piecewise-linear function:
10  % f(x)=-a*x+0.5(a+b)(|x+1|-|x-1|)
11  % For a folded torus chaotic attractor, try:
```

```
12   % [t,torusBreakdownOut]=torusBreakdown([0:0.1:1000],1e-5,1e-5,
13   % [0.1 0 0.1],15,1,0.07,0.1);
14
15   options = odeset('RelTol',reltol,'AbsTol',abstol);
16   [t,y] = ode45(@torusBreakdownFunction,tspan,x0,options);
17
18
19     function dy = torusBreakdownFunction(t,y)
20         dy = zeros(3,1); % a column vector
21         dy(1) = -alpha*f(y(2)-y(1),a,b);
22         dy(2) = -f(y(2)-y(1),a,b)-y(3);
23         dy(3) = beta*y(2);
24     end
25
26     function y=f(x,a,b)
27         y=-a*x+0.5*(a+b)*(abs(x+ones(length(x),1))-abs(x-ones(
               length(x),1)));
28     end
29   end
```

E.7 MATLAB Code for Quasi-Periodic Route to Chaos

Listing E.7 MATLAB code for obtaining torus-breakdown route to chaos in Fig. 4.4

```
1    % torus breakdown script
2    % make sure torusBreakdown.m is in the same folder
3    [t,torusAttractorTwoTorus]=torusBreakdown([0:0.1:1000],1e-5,1e
         -5,[0.1 0 0.1],2.0,1,0.07,0.1);
4    [t,torusAttractorPeriod8]=torusBreakdown([0:0.1:1000],1e-5,1e
         -5,[0.1 0 0.1],8.0,1,0.07,0.1);
5    [t,torusAttractorPeriod15]=torusBreakdown([0:0.1:1000],1e-5,1e
         -5,[0.1 0 0.1],8.8,1,0.07,0.1);
6    [t,torusAttractorTorusBreakdown]=torusBreakdown([0:0.1:1000],1e
         -5,1e-5,[0.1 0 0.1],15.0,1,0.07,0.1);
7
8    h1 = figure;
9    plot(torusAttractorTwoTorus([7500:10001],2),torusAttractorTwoTorus
         ([7500:10001],1))
10   % http://www.mathworks.com/matlabcentral/newsreader/view_thread
         /288159
11   a = get(gca,'XTickLabel');
12   set(gca,'XTickLabel',a,'fontsize',18)
13   xlabel('$y$','Interpreter','Latex','Fontsize',32)
14   ylabel('$x$','Interpreter','Latex','Fontsize',32)
15   title('Two-torus','Interpreter','Latex','Fontsize',32)
16   % enable line below for EPS output
17   %print(h1,'-depsc','-r600','chap4Figure-torusBreakDownTwoTorus.eps
         ');
18
19   h2=figure;
```

```matlab
20   plot(torusAttractorPeriod8([7500:10001],2),torusAttractorPeriod8
         ([7500:10001],1))
21   a = get(gca,'XTickLabel');
22   set(gca,'XTickLabel',a,'fontsize',18)
23   xlabel('$y$','Interpreter','Latex','Fontsize',32)
24   ylabel('$x$','Interpreter','Latex','Fontsize',32)
25   title('Period-8','Interpreter','Latex','Fontsize',32)
26   % enable line below for EPS output
27   %print(h2,'-depsc','-r600','chap4Figure-torusBreakDownPeriod8.eps
         ');

28
29   h3=figure;
30   plot(torusAttractorPeriod15([7500:10001],2),torusAttractorPeriod15
         ([7500:10001],1))
31   a = get(gca,'XTickLabel');
32   set(gca,'XTickLabel',a,'fontsize',18)
33   xlabel('$y$','Interpreter','Latex','Fontsize',32)
34   ylabel('$x$','Interpreter','Latex','Fontsize',32)
35   title('Period-15','Interpreter','Latex','Fontsize',32)
36   % enable line below for EPS output
37   %print(h3,'-depsc','-r600','chap4Figure-torusBreakDownPeriod15.eps
         ');

38
39   h4=figure;
40   plot(torusAttractorTorusBreakdown([7500:10001],2),
         torusAttractorTorusBreakdown([7500:10001],1))
41   a = get(gca,'XTickLabel');
42   set(gca,'XTickLabel',a,'fontsize',18)
43   xlabel('$y$','Interpreter','Latex','Fontsize',32)
44   ylabel('$x$','Interpreter','Latex','Fontsize',32)
45   title('Chaos','Interpreter','Latex','Fontsize',32)
46   % enable line below for EPS output
47   %print(h4,'-depsc','-r600','chap4Figure-
         torusAttractorTorusBreakdown.eps');
```

E.8 MATLAB Code with Chua Oscillator Parameter Values for Intermittency Route to Chaos

Listing E.8 MATLAB code implementing Eqs. (4.19)–(4.22)

```matlab
1   function [t,y] = chuaOscillatorIntermittency(tspan,reltol,
        abstol,x0,alpha,beta,gamma,a,b)
2
3   % Simulates Chua's oscillator with a piecewise-linear
        nonlinearity
4   % x'=alpha*(y-x-f(x))
5   % y'=x-y+z
6   % z'=-beta*y-gamma*z
7   % f(x) = bx + 0.5*(a-b)*(|x+1|-|x-1|)
8   % Function uses ode45. The arguments to be passed into the
        function
```

```
9   % are tspan, reltol,abstol,x0,alpha,beta,,gamma,a and b.
10
11  options = odeset('RelTol',reltol,'AbsTol',abstol);
12  [t,y] = ode45(@chuaOscillatorFunction,tspan,x0,options);
13
14
15      function dy = chuaOscillatorFunction(t,y)
16          dy = zeros(3,1); % a column vector
17          dy(1) = alpha*(y(2)-y(1)-f(y(1),a,b));
18          dy(2) = y(1)-y(2)+y(3);
19          dy(3) = -beta*y(2)-gamma*y(3);
20      end
21
22      function y=f(x,a,b)
23          y=b*x+0.5*(a-b)*(abs(x+ones(length(x),1))-abs(x-ones(
                length(x),1)));
24      end
25  end
```

E.9 MATLAB Code for Plotting Intermittency Route to Chaos

Listing E.9 MATLAB code for obtaining intermittency route to chaos in Fig. 4.5

```
1   % Intermittency script
2   % make sure chuaOscillatorIntermittency.m is in the same folder
3   alpha=-75.018755;
4   a=-0.98;
5   b=-2.4;
6
7   % Periodic
8   beta=44.803;
9   gamma=-4.480;
10  [t,intermittencyPeriodic]=chuaOscillatorIntermittency([0:0.01:1000],1e
        -4,1e-4,[0.1 0 0.1],alpha,beta,gamma,a,b);
11  h1a = figure;
12  plot(intermittencyPeriodic(:,2),intermittencyPeriodic(:,3))
13  % http://www.mathworks.com/matlabcentral/newsreader/view_thread/288159
14  figureProperties = get(gca,'XTickLabel');
15  set(gca,'XTickLabel',figureProperties,'fontsize',18)
16  xlabel('$y$','Interpreter','Latex','Fontsize',32)
17  ylabel('$z$','Interpreter','Latex','Fontsize',32)
18  title('Limit Cycle(s)','Interpreter','Latex','Fontsize',32)
19  % enable line below for EPS output
20  % print(h1a,'-depsc','-r600','chap4Figure-intermittencyPeriodic.eps');
21
22  h1b = figure;
23  plot(t([75000:83000],1),intermittencyPeriodic([75000:83000],1));
24  % http://www.mathworks.com/matlabcentral/newsreader/view_thread/288159
25  figureProperties = get(gca,'XTickLabel');
26  set(gca,'XTickLabel',figureProperties,'fontsize',18)
27  xlabel('$t$','Interpreter','Latex','Fontsize',32)
28  ylabel('$x$','Interpreter','Latex','Fontsize',32)
```

```
29   title('Limit Cycle(s) (time-domain)','Interpreter','Latex','Fontsize'
         ,32)
30   %enable line below for EPS output
31   % print(h1b,'-depsc','-r600','chap4Figure-intermittencyPeriodic-
         TimeDomain.eps');
32
33   % Chaos, one instance
34   beta=43.994721;
35   gamma=-4.3994721;
36   [t,intermittencyChaosOne]=chuaOscillatorIntermittency([0:0.01:1000],1e
         -6,1e-6,[0.1,0,0.1],alpha,beta,gamma,a,b);
37   h2a = figure;
38   plot(intermittencyChaosOne([75000:100001],3),intermittencyChaosOne
         ([75000:100001],1))
39   % http://www.mathworks.com/matlabcentral/newsreader/view_thread/288159
40   figureProperties = get(gca,'XTickLabel');
41   set(gca,'XTickLabel',figureProperties,'fontsize',18)
42   xlabel('$z$','Interpreter','Latex','Fontsize',32)
43   ylabel('$x$','Interpreter','Latex','Fontsize',32)
44   title('Intermittency','Interpreter','Latex','Fontsize',32)
45   % enable line below for EPS output
46   % print(h2a,'-depsc','-r600','chap4Figure-intermittencyChaosOne.eps');
47
48   h2b = figure;
49   plot(t([75000:83000],1),intermittencyChaosOne([75000:83000],2))
50   % http://www.mathworks.com/matlabcentral/newsreader/view_thread/288159
51   figureProperties = get(gca,'XTickLabel');
52   set(gca,'XTickLabel',figureProperties,'fontsize',18)
53   xlabel('$t$','Interpreter','Latex','Fontsize',32)
54   ylabel('$y$','Interpreter','Latex','Fontsize',32)
55   title('Intermittency Chaos (time-domain)','Interpreter','Latex','
         Fontsize',32)
56   % enable line below for EPS output
57   % print(h2b,'-depsc','-r600','chap4Figure-intermittencyChaosOne-
         TimeDomain.eps');
58
59   % Chaos, second instance
60   beta=31.746032;
61   gamma=-3.1746032;
62   [t,intermittencyChaosTwo]=chuaOscillatorIntermittency([0:0.01:1000],1e
         -6,1e-6,[0.1,0,0.1],alpha,beta,gamma,a,b);
63   h3a = figure;
64   plot(intermittencyChaosTwo([75000:100001],3),intermittencyChaosTwo
         ([75000:100001],1))
65   % http://www.mathworks.com/matlabcentral/newsreader/view_thread/288159
66   figureProperties = get(gca,'XTickLabel');
67   set(gca,'XTickLabel',figureProperties,'fontsize',18)
68   xlabel('$z$','Interpreter','Latex','Fontsize',32)
69   ylabel('$x$','Interpreter','Latex','Fontsize',32)
70   title('Intermittency','Interpreter','Latex','Fontsize',32)
71   % enable line below for EPS output
72   % print(h3a,'-depsc','-r600','chap4Figure-intermittencyChaosTwo.eps');
73
74   h3b = figure;
75   plot(t([75000:83000],1),intermittencyChaosTwo([75000:83000],2))
76   % http://www.mathworks.com/matlabcentral/newsreader/view_thread/288159
77   figureProperties = get(gca,'XTickLabel');
78   set(gca,'XTickLabel',figureProperties,'fontsize',18)
79   xlabel('$t$','Interpreter','Latex','Fontsize',32)
```

```matlab
80  ylabel('$y$','Interpreter','Latex','Fontsize',32)
81  title('Intermittency Chaos (time-domain)','Interpreter','Latex','
        Fontsize',32)
82  % enable line below for EPS output
83  % print(h3b,'-depsc','-r600','chap4Figure-intermittencyChaosTwo-
        TimeDomain.eps');

85  % Chaos, third instance
86  beta=31.25;
87  gamma=-3.125;
88  [t,intermittencyChaosThree]=chuaOscillatorIntermittency([0:0.01:1000],1e
        -6,1e-6,[0.1,0,0.1],alpha,beta,gamma,a,b);
89  h4a = figure;
90  plot(intermittencyChaosThree([75000:100001],3),intermittencyChaosThree
        ([75000:100001],1))
91  % http://www.mathworks.com/matlabcentral/newsreader/view_thread/288159
92  figureProperties = get(gca,'XTickLabel');
93  set(gca,'XTickLabel',figureProperties,'fontsize',18)
94  xlabel('$z$','Interpreter','Latex','Fontsize',32)
95  ylabel('$x$','Interpreter','Latex','Fontsize',32)
96  title('Intermittency','Interpreter','Latex','Fontsize',32)
97  % enable line below for EPS output
98  % print(h4a,'-depsc','-r600','chap4Figure-intermittencyChaosThree.eps');

100 h4b = figure;
101 plot(t([75000:83000],1),intermittencyChaosThree([75000:83000],2))
102 % http://www.mathworks.com/matlabcentral/newsreader/view_thread/288159
103 figureProperties = get(gca,'XTickLabel');
104 set(gca,'XTickLabel',figureProperties,'fontsize',18)
105 xlabel('$t$','Interpreter','Latex','Fontsize',32)
106 ylabel('$y$','Interpreter','Latex','Fontsize',32)
107 title('Intermittency Chaos (time-domain)','Interpreter','Latex','
        Fontsize',32)
108 % enable line below for EPS output
109 % print(h4b,'-depsc','-r600','chap4Figure-intermittencyChaosThree-
        TimeDomain.eps');
```

E.10 MATLAB Code for Resource-Consumer-Predator Model

Listing E.10 MATLAB code implementing Eqs. (4.23)–(4.25)

```matlab
1   function [t,y] = resourcePredatorPrey(tspan,reltol,abstol,x0,xC,
        yC,xP,yP,R0,C0,K)

3   % Simulates the resource predator prey model from
4   % "Controlling transient chaos in deterministic flows with
        applications
5   % to electrical power systems and ecology". Physical Review E,
        59(2),
6   % 1646 - 1655, 1999.
7   % Function uses ode45. The arguments to be passed into the
        function
8   % are tspan, reltol,abstol,x0,alpha,beta,,gamma,a and b.
```

```
 9
10    options = odeset('RelTol',reltol,'AbsTol',abstol);
11    [t,y] = ode45(@resourcePredatorPreyFunction,tspan,x0,options);
12        function dy = resourcePredatorPreyFunction(t,y)
13            dy = zeros(3,1); % a column vector
14            dy(1) = y(1)*(1-y(1)/K)-(xC*yC*y(2)*y(1))/(y(1)+R0);
15            dy(2) = xC*y(2)*((yC*y(1))/(y(1)+R0)-1)-(xP*yP*y(3)*y(2))/(
                    y(2)+C0);
16            dy(3) = xP*y(3)*(-1+(yP*y(2))/(y(2)+C0));
17        end
18
19
20    end
```

E.11 MATLAB Code for Chaotic Transients

Listing E.11 MATLAB code for simulating chaotic transients and crisis phenomenon in Fig. 4.6

```
 1    % chaotic transients script
 2    % make sure resourcePredatorPrey.m is in the same folder
 3    xC=0.4;
 4    yC=2.009;
 5    xP=0.08;
 6    yP=2.876;
 7    R0=0.16129;
 8    C0=0.5;
 9
10    % Periodic attractor and chaotic attractor co-exist
11    K=0.99;
12    [t,chaoticTransientPeriodic]=resourcePredatorPrey
          ([0:0.1:1000],1e-4,1e-4,[0.1 0.2 0.1],xC,yC,xP,yP,R0,C0,
          K);
13    h1 = figure;
14    plot(chaoticTransientPeriodic(:,1),chaoticTransientPeriodic
          (:,2))
15    % http://www.mathworks.com/matlabcentral/newsreader/
          view_thread/288159
16    figureProperties = get(gca,'XTickLabel');
17    set(gca,'XTickLabel',figureProperties,'fontsize',18)
18    xlabel('$R$','Interpreter','Latex','Fontsize',32)
19    ylabel('$C$','Interpreter','Latex','Fontsize',32)
20    title('Limit Cycle','Interpreter','Latex','Fontsize',32)
21    % enable line below for EPS output
22    % print(h1,'-depsc','-r600','chap4Figure-
          chaoticTransientsPeriodic.eps');
23
24    h2 = figure;
25    plot(t(:,1),chaoticTransientPeriodic(:,3))
```

```
26  % http://www.mathworks.com/matlabcentral/newsreader/
        view_thread/288159
27  figureProperties = get(gca,'XTickLabel');
28  set(gca,'XTickLabel',figureProperties,'fontsize',18)
29  xlabel('$t$','Interpreter','Latex','Fontsize',32)
30  ylabel('$P$','Interpreter','Latex','Fontsize',32)
31  title('$P$ population decays','Interpreter','Latex','
        Fontsize',32)
32  %enable line below for EPS output
33  % print(h2,'-depsc','-r600','chap4Figure-
        chaoticTransientsPeriodic-TimeDomain.eps');
34
35  h3 = figure;
36  [t,chaoticTransientChaos]=resourcePredatorPrey
        ([0:0.1:1000],1e-4,1e-4,[0.55 0.35 0.8],xC,yC,xP,yP,R0,
        C0,K);
37  plot(chaoticTransientChaos(:,2),chaoticTransientChaos(:,3))
38  % http://www.mathworks.com/matlabcentral/newsreader/
        view_thread/288159
39  figureProperties = get(gca,'XTickLabel');
40  set(gca,'XTickLabel',figureProperties,'fontsize',18)
41  xlabel('$R$','Interpreter','Latex','Fontsize',32)
42  ylabel('$P$','Interpreter','Latex','Fontsize',32)
43  title('Chaos','Interpreter','Latex','Fontsize',32)
44  %enable line below for EPS output
45  % print(h3,'-depsc','-r600','chap4Figure-
        chaoticTransientsChaos.eps');
46
47  % Crisis past critical carrying capacity.
48  K=1.02;
49  [t,chaoticTransientCrisis]=resourcePredatorPrey
        ([0:0.1:1000],1e-4,1e-4,[0.55 0.35 0.8],xC,yC,xP,yP,R0,
        C0,K);
50  h4 = figure;
51  plot(t(:,1),chaoticTransientCrisis(:,3))
52  % http://www.mathworks.com/matlabcentral/newsreader/
        view_thread/288159
53  figureProperties = get(gca,'XTickLabel');
54  set(gca,'XTickLabel',figureProperties,'fontsize',18)
55  xlabel('$t$','Interpreter','Latex','Fontsize',32)
56  ylabel('$P$','Interpreter','Latex','Fontsize',32)
57  title('$P$ time series','Interpreter','Latex','Fontsize',32)
58  % enable line below for EPS output
59  % print(h4,'-depsc','-r600','chap4Figure-
        chaoticTransientsCrisis.eps');
```

E.12   VHDL Specification for Single Pulse Generator

Listing E.12   VHDL pulse FSM

```vhdl
 1  library ieee;
 2  use ieee.std_logic_1164.all;
 3  use ieee.numeric_std.all;
 4
 5  entity pulseFSM is port (
 6        reset,clock,trigger : in std_logic;
 7        pulseOut,pulseOutSingleClockCycle : out std_logic);
 8  end pulseFSM;
 9
10  architecture mooreFSM of pulseFSM is
11
12      type state is (resetState,generatePulseCycle1,
13          generatePulseCycle2,generatePulseCycle3,
            generatePulseCycle4,generatePulseCycle5,
            generatePulseCycle6,generatePulseCycle7,
            generatePulseCycle8,stopPulse,waitForTriggerRelease);
        signal currentState,nextState : state;
14
15  begin
16
17      -- state memory
18      stateMemory : process(reset,clock)
19      begin
20        if reset='1' then
21           currentState <= resetState;
22        else
23           if rising_edge(clock) then
24              currentState <= nextState;
25           end if;
26        end if;
27      end process;
28
29      -- next state logic
30      stateTransitionLogic : process (currentState,trigger)
31      begin
32        case currentState is
33           when resetState =>
34                 if trigger='0' then
35                    nextState <= resetState;
36                 else
37                    nextState <= generatePulseCycle1;
38                 end if;
39           when generatePulseCycle1 =>
40                    nextState <= generatePulseCycle2;
41           when generatePulseCycle2 =>
42                    nextState <= generatePulseCycle3;
43           when generatePulseCycle3 =>
```

```
44                          nextState <= generatePulseCycle4;
45              when generatePulseCycle4 =>
46                          nextState <= generatePulseCycle5;
47              when generatePulseCycle5 =>
48                          nextState <= generatePulseCycle6;
49              when generatePulseCycle6 =>
50                          nextState <= generatePulseCycle7;
51              when generatePulseCycle7 =>
52                          nextState <= generatePulseCycle8;
53              when generatePulseCycle8 =>
54                          nextState <= stopPulse;
55              when stopPulse =>
56                          nextState <= waitForTriggerRelease;
57              when waitForTriggerRelease =>
58                      if trigger='1' then
59                          nextState <= waitForTriggerRelease;
60                      else
61                          nextState <= resetState;
62                      end if;
63          end case;
64      end process;
65
66      -- output logic
67      with currentState select
68          pulseOut <= '0' when resetState,
69                          '0' when waitForTriggerRelease,
70                          '0' when stopPulse,
71                          '1' when others;
72
73      -- we enable single clock cycle pulse only after latency
                of floating point design has been
74      -- accounted for.
75      with currentState select
76          pulseOutSingleClockCycle <= '1' when stopPulse,
77                          '0' when others;
78  end mooreFSM;
```

E.13 ModelSim Testbench for Single Pulse Generator

Listing E.13 VHDL pulse FSM test bench

```
1   -- testbench for pulse FSM (bifurcations)
2
3   library ieee;
4   use ieee.std_logic_1164.all;
5   use ieee.numeric_std.all;
6
7   entity pulseFSMtb is
8   end pulseFSMtb;
```

```
 9
10  architecture testbench of pulseFSMtb is
11     signal clock,reset,trigger,pulseOut : std_logic := '0';
12
13     component pulseFSM is port (
14        reset,clock,trigger : in std_logic;
15        pulseOut : out std_logic);
16     end component;
17
18  begin
19     pulseFSMInstance : pulseFSM port map (
20        reset => reset,
21        clock => clock,
22        trigger => trigger,
23        pulseOut => pulseOut);
24
25     clock50MHzProcess : process
26     begin
27        clock <= not clock;
28        wait for 10 ns;
29     end process clock50MHzProcess;
30
31     stimulus : process
32     begin
33        reset <= '1';
34        trigger <= '0';
35        wait for 55 ns;
36        reset <= '0'; -- unreset after 55 ns
37        wait for 100 ns;
38        trigger <= '1';
39        wait for 150 ns;
40        trigger <= '0';
41        wait; -- prevent process from being executed again
42     end process stimulus;
43  end testbench;
```

E.14 ModelSim Script File for Single Pulse Generator

Listing E.14 ModelSim script file for pulse FSM

```
 1  # do vlib work only once!
 2  # vlib work
 3
 4  vmap work work
 5
 6  # compile source
 7  vcom ../../pulseFSM.vhd
 8
 9  # compile testbench
10  vcom pulseFSMtb.vhd
11
```

```
12   vsim pulseFSMtb
13   # configure wave window to have a white background color
14   # http://www.utdallas.edu/~zhoud/EE%203120/
         Xilinx_tutorial_Spartan3_home_PC.pdf and ModelSim
         Reference Manual - configure command
15   configure wave -background white -foreground red -textcolor
         blue -timecolor blue -vectorcolor red -wavebackground
         white -cursorcolor black
16   add wave -divider "Clock and Reset"
17   add wave clock
18   add wave reset
19
20   add wave -divider "Input"
21   add wave trigger
22   add wave -divider "Output"
23   add wave pulseOut
24   add wave -divider "FSM states"
25   add wave -label "Synchronous current state" sim:/pulsefsmtb/
         pulseFSMInstance/currentState
26
27   # run 500ns
```

E.15   VHDL Specification of Period-Doubling Route to Chaos

Listing E.15 VHDL specification of period-doubling bifurcation in the Rössler system

```
1   --Single precision (32-bit) floating point realization
2
3   library ieee;
4   use ieee.std_logic_1164.all;
5   use ieee.std_logic_signed.all;
6
7   entity rosslerSystem is port (
8       resetn, clockIn, incrementCountClockN,
           incrementGammaClockN,incrementCount,incrementGamma :
           in std_logic;
9       xOut,yOut,zOut : out std_logic_vector(15 downto 0));
10  end rosslerSystem;
11
12  architecture behavioral of rosslerSystem is
13
14  signal reset,incrementGammaClock,incrementDecrementGammaPulse,
         dFlipFlopClock : std_logic;
15  -- constants
16  signal dt,output1Over2Scale,output1Over5Scale,alpha,beta,gamma,
         gammaSignal : std_logic_vector(31 downto 0);
17  -- state variables
18  signal x,y,z,xNew,yNew,zNew,xScaled,yScaled,zScaled,xFixed,
         yFixed,zFixed : std_logic_vector(31 downto 0);
19  -- prescalar
20  signal count,countIncrement : integer range 0 to 128;
```

```vhdl
21
22  -- DSP builder top level.
23  -- Note: Open rosslerSystem_NonlinearSubsystem.vhd and create
            component.
24  -- NOTE : SAME STEPS FOR OTHER CHAOTIC SYSTEMS!
25  component rosslerSystem_NonlinearSubsystem is
26      port (
27          In_a : in std_logic_vector(31 downto 0);
28          In_b : in std_logic_vector(31 downto 0);
29          In_dt : in std_logic_vector(31 downto 0);
30          In_g : in std_logic_vector(31 downto 0);
31          In_x : in std_logic_vector(31 downto 0);
32          In_y : in std_logic_vector(31 downto 0);
33          In_z : in std_logic_vector(31 downto 0);
34          Out_x : out std_logic_vector(31 downto 0);
35          Out_y : out std_logic_vector(31 downto 0);
36          Out_z : out std_logic_vector(31 downto 0);
37          clk : in std_logic;
38          areset : in std_logic;
39          h_areset : in std_logic
40          );
41  end component;
42  -- END DSP builder top level.
43
44  -- latency : 5 clock cycles (scale for DAC range)
45  component floatingPointMultiplyDedicated IS
46      PORT
47      (
48          aclr : IN STD_LOGIC ;
49          clock : IN STD_LOGIC ;
50          dataa : IN STD_LOGIC_VECTOR (31 DOWNTO 0);
51          datab : IN STD_LOGIC_VECTOR (31 DOWNTO 0);
52          result : OUT STD_LOGIC_VECTOR (31 DOWNTO 0)
53      );
54  END component;
55
56  -- latency : 6 clock cycles
57  component floatingPointToFixed IS
58      PORT
59      (
60          aclr : IN STD_LOGIC ;
61          clock : IN STD_LOGIC ;
62          dataa : IN STD_LOGIC_VECTOR (31 DOWNTO 0);
63          result : OUT STD_LOGIC_VECTOR (31 DOWNTO 0)
64      );
65  END component;
66
67  component floatingPointAddSubtract IS
68      PORT
69      (
70          aclr : IN STD_LOGIC ;
71          add_sub : IN STD_LOGIC ;
72          clk_en : IN STD_LOGIC ;
```

```
73          clock : IN STD_LOGIC ;
74          dataa : IN STD_LOGIC_VECTOR (31 DOWNTO 0);
75          datab : IN STD_LOGIC_VECTOR (31 DOWNTO 0);
76          result : OUT STD_LOGIC_VECTOR (31 DOWNTO 0)
77      );
78   END component;
79
80   component pulseFSM is port (
81          reset,clock,trigger : in std_logic;
82          pulseOut,pulseOutSingleClockCycle : out std_logic);
83   end component;
84
85   component dFlipFlopWithAsyncReset is port (
86      clock,reset : in std_logic;
87      d,resetVal : in std_logic_vector(31 downto 0);
88      q : out std_logic_vector(31 downto 0));
89   end component;
90
91   begin
92      reset <= not resetn;
93      -- Euler's method
94      -- We first synchronously update state variables at 781.250
                KHz (64 counts of 50 MHz clock)
95      -- Since dt = 1/1024, time scale is actually (780.250e3
                /1024) = 762 Hz (approximately)
96
97      -- since synchronous update count is integer, simply use a
                process statement
98      process(incrementCountClockN,resetn)
99      begin
100         if resetn = '0' then
101             countIncrement <= 64;
102         else
103            if falling_edge(incrementCountClockN) then
104                if incrementCount = '1' then
105                    countIncrement <= countIncrement+1;
106                else
107                    countIncrement <= countIncrement-1;
108                end if;
109            end if;
110         end if;
111      end process;
112      -- state memory
113      process(clockIn, resetn)
114      begin
115             -- constants (place outside reset and clock to avoid
                    latches)
116             dt <= X"3A800000"; -- 1/1024
117             output1Over5Scale <= X"3E4CCCCC";
118             output1Over2Scale <= X"3F000000";
119             -- default values for parameters and synchronous
                    count
120             alpha <= X"3DCCCCCC"; --0.1
```

```
121                    beta <= X"3DCCCCCC"; --0.1
122                if resetn = '0' then
123                    -- initial state
124                    x <= X"41680000"; -- 14.5
125                    y <= X"00000000"; -- 0
126                    z <= X"3DCCCCCC"; -- 0.1
127                    count <= 0;
128                else
129                    if rising_edge(clockIn) then
130                        if count = countIncrement then
131                            count <= 0;
132                        else
133                            count <= count + 1;
134                        end if;
135
136                        if count = countIncrement-1 then
137                            x <= xNew;
138                            y <= yNew;
139                            z <= zNew;
140                        end if;
141                    end if;
142            end if;
143        end process;
144
145        incrementGammaClock <= not incrementGammaClockN;
146        pulseFSMForGamma : pulseFSM port map (
147            reset => reset,
148            clock => clockIn,
149            trigger => incrementGammaClock,
150            pulseOut => incrementDecrementGammaPulse,
151            pulseOutSingleClockCycle => dFlipFlopClock);
152
153        gammaParameterBifurcation : floatingPointAddSubtract port
             map (
154            aclr => reset,
155            add_sub => incrementGamma, -- '1' = add, '0' = subtract
156            clk_en => incrementDecrementGammaPulse,
157            clock => clockIn,
158            dataa => gamma, -- start at 12 = X"41400000". Memory
                 implemented using D flip-flop
159            datab => X"3DCCCCCC", -- increment/decrement by 0.1 = X"3
                 DCCCCCC"
160            result => gammaSignal);
161
162    -- we will need the flip-flop below to provide a proper initial
           state.
163    gammaFlipFlop : dFlipFlopWithAsyncReset port map (
164        clock => dFlipFlopClock,
165        reset => reset,
166        d => gammaSignal,
167        resetVal => X"41400000", -- start at 12
168        q => gamma);
169
```

```
170     staticNonlinearitiesAndDeltaT :
            rosslerSystem_NonlinearSubsystem port map (
171         In_a => alpha,
172         In_b => beta,
173         In_dt => dt,
174         In_g => gamma,
175           In_x => x,
176         In_y => y,
177         In_z => z,
178         Out_x => xNew,
179         Out_y => yNew,
180         Out_z => zNew,
181         clk => clockIn,
182         areset => reset,
183         h_areset => reset);
184     -- END Euler's method
185
186     -- scale outputs
187     scaleX : floatingPointMultiplyDedicated port map (
188         aclr => reset,
189         clock => clockIn,
190         dataa => x,
191         datab => output1Over2Scale,
192         result => xScaled);
193     scaleY : floatingPointMultiplyDedicated port map (
194         aclr => reset,
195         clock => clockIn,
196         dataa => y,
197         datab => output1Over2Scale,
198         result => yScaled);
199     scaleZ : floatingPointMultiplyDedicated port map (
200         aclr => reset,
201         clock => clockIn,
202         dataa => z,
203         datab => output1Over5Scale,
204         result => zScaled);
205
206     --state outputs : convert scaled floating point x,y
                variables to 2.30 fixed point for DAC
207     xOutFinal : floatingPointToFixed port map (
208         aclr => reset,
209         clock => clockIn,
210         dataa => xScaled,
211         result => xFixed);
212     yOutFinal : floatingPointToFixed port map (
213         aclr => reset,
214         clock => clockIn,
215         dataa => yScaled,
216         result => yFixed);
217     zOutFinal : floatingPointToFixed port map (
218         aclr => reset,
219         clock => clockIn,
220         dataa => zScaled,
```

```
221            result => zFixed);
222
223            xOut <= xFixed(31 downto 16);
224            yOut <= yFixed(31 downto 16);
225            zOut <= zFixed(31 downto 16);
226
227    end behavioral;
```

1. Lines 67–89 show how we have decided to implement the bifurcation scenario. Instead of using DSP builder to increment (decrement) a parameter, we have utilized the MegaWizard. The primary reason is that the increment or decrement of the bifurcation parameter simply requires only one module: floating point addition and subtraction.

2. Lines 145–168 show how we implement the bifurcation. We generate two single-cycle 20 ns wide clock pulses using the pulse FSM. The first clock pulse is used as a clock enable signal for the appropriate floating point module that implements the bifurcation sequence. In this example, we use the floating-point addsub module (lines 153–160) to specify gamma as the bifurcation parameter. The second clock pulse is delayed by the appropriate number of clock cycles (depending on the floating point module) and used to clock the bifurcation parameter register. In this design, we have a delay of nine clock cycles due to the eight clock cycle latency associated with the floating-point addsub module.

E.16 VHDL Specification for Period-Adding Route to Chaos

Listing E.16 VHDL specification of period-adding bifurcation in the Chua system

```
1    --Single precision (32-bit) floating point realization
2    library ieee;
3    use ieee.std_logic_1164.all;
4    use ieee.std_logic_signed.all;
5
6    entity chuaOscillator is port (
7          resetn, clockIn, incrementCountPulseN,
                incrementBetaPulseN, incrementCount, incrementBeta :
                in std_logic;
8          xOut, yOut, zOut : out std_logic_vector(15 downto 0));
9    end chuaOscillator;
10
11   architecture behavioral of chuaOscillator is
12
13   signal reset, incrementBetaPulse, incrementDecrementBetaPulse,
          dFlipFlopClock : std_logic;
14   -- constants
15   signal dt, outputScaledBy8, alpha, beta, betaSignal, gamma, a, c :
          std_logic_vector(31 downto 0);
16   -- state variables
17   signal x, y, z, xNew, yNew, zNew, xScaled, yScaled, zScaled, xFixed,
          yFixed, zFixed : std_logic_vector(31 downto 0);
```

```vhdl
18   -- prescalar
19   signal count,countIncrement : integer range 0 to 128;
20
21   -- DSP builder top level.
22   component chuaOscillator_NonlinearSubsystem is
23     port (
24         In_a : in std_logic_vector(31 downto 0);
25         In_alpha : in std_logic_vector(31 downto 0);
26         In_beta : in std_logic_vector(31 downto 0);
27         In_c : in std_logic_vector(31 downto 0);
28         In_dt : in std_logic_vector(31 downto 0);
29         In_gamma : in std_logic_vector(31 downto 0);
30         In_x : in std_logic_vector(31 downto 0);
31         In_y : in std_logic_vector(31 downto 0);
32         In_z : in std_logic_vector(31 downto 0);
33         Out_x : out std_logic_vector(31 downto 0);
34         Out_y : out std_logic_vector(31 downto 0);
35         Out_z : out std_logic_vector(31 downto 0);
36         clk : in std_logic;
37         areset : in std_logic;
38         h_areset : in std_logic);
39   end component;
40   -- END DSP builder top level.
41
42   -- latency : 5 clock cycles (scale for DAC range)
43   component floatingPointMultiplyDedicated IS
44     PORT
45     (
46         aclr : IN STD_LOGIC ;
47         clock : IN STD_LOGIC ;
48         dataa : IN STD_LOGIC_VECTOR (31 DOWNTO 0);
49         datab : IN STD_LOGIC_VECTOR (31 DOWNTO 0);
50         result : OUT STD_LOGIC_VECTOR (31 DOWNTO 0)
51     );
52   END component;
53
54   -- latency : 6 clock cycles
55   component floatingPointToFixed IS
56     PORT
57     (
58         aclr : IN STD_LOGIC ;
59         clock : IN STD_LOGIC ;
60         dataa : IN STD_LOGIC_VECTOR (31 DOWNTO 0);
61         result : OUT STD_LOGIC_VECTOR (31 DOWNTO 0)
62     );
63   END component;
64
65   component floatingPointAddSubtract IS
66     PORT
67     (
68         aclr : IN STD_LOGIC ;
69         add_sub : IN STD_LOGIC ;
70         clk_en : IN STD_LOGIC ;
```

```vhdl
71          clock : IN STD_LOGIC ;
72          dataa : IN STD_LOGIC_VECTOR (31 DOWNTO 0);
73          datab : IN STD_LOGIC_VECTOR (31 DOWNTO 0);
74          result : OUT STD_LOGIC_VECTOR (31 DOWNTO 0)
75      );
76  END component;
77
78  component pulseFSM is port (
79      reset,clock,trigger : in std_logic;
80      pulseOut,pulseOutSingleClockCycle : out std_logic);
81  end component;
82
83  component dFlipFlopWithAsyncReset is port (
84      clock,reset : in std_logic;
85      d,resetVal : in std_logic_vector(31 downto 0);
86      q : out std_logic_vector(31 downto 0));
87  end component;
88
89  begin
90      reset <= not resetn;
91      -- Euler's method
92      -- We first synchronously update state variables at
           781.250 KHz (64 counts of 50 MHz clock)
93      -- Since dt = 1/1024, time scale is actually (780.250e3
           /1024) = 762 Hz (approximately)
94
95      -- since synchronous update count is integer, simply use
           a process statement
96      process(incrementCountPulseN,resetn)
97      begin
98          if resetn = '0' then
99              countIncrement <= 64;
100         else
101             if falling_edge(incrementCountPulseN) then
102                 if incrementCount = '1' then
103                     countIncrement <= countIncrement+1;
104                 else
105                     countIncrement <= countIncrement-1;
106                 end if;
107             end if;
108         end if;
109     end process;
110     -- state memory
111     process(clockIn, resetn)
112     begin
113             -- constants (place outside reset and clock to
                   avoid latches)
114             dt <= X"3A800000"; -- 1/1024
115             outputScaledBy8 <= X"41000000";
116             -- default values for parameters and synchronous
                   count
117             alpha <= X"406D4FDF"; -- 3.708, approximately
                   3.70799999
```

```
118            beta <= X"40666666"; -- 3.6, approximately
                  3.5999999
119            gamma <= X"3D9BA5E3"; -- 0.076, approximately
                  0.0759999
120            a <= X"3F800000"; -- 1
121            c <= X"BE8D4FDF"; -- -0.276, approximately
                  -0.27599999
122        if resetn = '0' then
123            -- initial state
124            x <= X"3DCCCCCC"; -- 0.1
125            y <= X"00000000"; -- 0
126            z <= X"3DCCCCCC"; -- 0.1
127            count <= 0;
128        else
129            if rising_edge(clockIn) then
130                if count = countIncrement then
131                    count <= 0;
132                else
133                    count <= count + 1;
134                end if;
135
136                if count = countIncrement-1 then
137                    x <= xNew;
138                    y <= yNew;
139                    z <= zNew;
140                end if;
141            end if;
142        end if;
143    end process;
144
145 -- incrementBetaPulse <= not incrementBetaPulseN;
146 -- pulseFSMForBeta : pulseFSM port map (
147 -- reset => reset,
148 -- clock => clockIn,
149 -- trigger => incrementBetaPulse,
150 -- pulseOut => incrementDecrementBetaPulse,
151 -- pulseOutSingleClockCycle => dFlipFlopClock);
152 --
153 -- betaParameterBifurcation : floatingPointAddSubtract port
       map (
154 -- aclr => reset,
155 -- add_sub => incrementBeta, -- '1' = add, '0' = subtract
156 -- clk_en => incrementDecrementBetaPulse,
157 -- clock => clockIn,
158 -- dataa => beta, -- start at ????????. Memory implemented
       using D flip-flop
159 -- datab => X"", -- increment/decrement by ???? = X
       "??????????"
160 -- result => betaSignal);
161 --
162 ---- we will need the flip-flop below to provide a proper
       initial state.
163 -- betaFlipFlop : dFlipFlopWithAsyncReset port map (
```

```
164   -- clock => dFlipFlopClock,
165   -- reset => reset,
166   -- d => betaSignal,
167   -- resetVal => X"", -- start at ??????????
168   -- q => beta);
169   --
170      staticNonlinearitiesAndDeltaT :
            chuaOscillator_NonlinearSubsystem port map (
171         In_a => a,
172         In_alpha => alpha,
173         In_beta => beta,
174         In_c => c,
175         In_dt => dt,
176         In_gamma => gamma,
177         In_x => x,
178         In_y => y,
179         In_z => z,
180         Out_x => xNew,
181         Out_y => yNew,
182         Out_z => zNew,
183         clk => clockIn,
184         areset => reset,
185         h_areset => reset);
186      -- END Euler's method
187
188      -- scale outputs
189      scaleX : floatingPointMultiplyDedicated port map (
190         aclr => reset,
191         clock => clockIn,
192         dataa => x,
193         datab => outputScaledBy8,
194         result => xScaled);
195      scaleY : floatingPointMultiplyDedicated port map (
196         aclr => reset,
197         clock => clockIn,
198         dataa => y,
199         datab => outputScaledBy8,
200         result => yScaled);
201      scaleZ : floatingPointMultiplyDedicated port map (
202         aclr => reset,
203         clock => clockIn,
204         dataa => z,
205         datab => outputScaledBy8,
206         result => zScaled);
207
208      --state outputs : convert scaled floating point x,y
            variables to 2.30 fixed point for DAC
209      xOutFinal : floatingPointToFixed port map (
210         aclr => reset,
211         clock => clockIn,
212         dataa => xScaled,
213         result => xFixed);
214      yOutFinal : floatingPointToFixed port map (
```

```
215        aclr => reset,
216        clock => clockIn,
217        dataa => yScaled,
218        result => yFixed);
219     zCutFinal : floatingPointToFixed port map (
220        aclr => reset,
221        clock => clockIn,
222        dataa => zScaled,
223        result => zFixed);
224
225        xOut <= xFixed(31 downto 16);
226        yOut <= yFixed(31 downto 16);
227        zOut <= zFixed(31 downto 16);
228
229  end behavioral;
```

E.17 VHDL Specification for Quasi-Periodic Route to Chaos

Listing E.17 VHDL specification of quasi-periodic route to chaos via torus-breakdown

```
1   --Single precision (32-bit) floating point realization
2   library ieee;
3   use ieee.std_logic_1164.all;
4   use ieee.std_logic_signed.all;
5
6   entity torusBreakdown is port (
7        resetn, clockIn, incrementCountPulseN,
             incrementAlphaPulseN, incrementCount, incrementAlpha
             : in std_logic;
8        xOut, yOut, zOut : out std_logic_vector(15 downto 0));
9   end torusBreakdown;
10
11  architecture behavioral of torusBreakdown is
12
13  signal reset, incrementAlphaPulse,
         incrementDecrementAlphaPulse, dFlipFlopClock : std_logic;
14  -- constants
15  signal dt, outputScaledBy4, alpha, beta, alphaSignal, a, b :
         std_logic_vector(31 downto 0);
16  -- state variables
17  signal x, y, z, xNew, yNew, zNew, xScaled, yScaled, zScaled, xFixed,
         yFixed, zFixed : std_logic_vector(31 downto 0);
18  -- prescalar
19  signal count, countIncrement : integer range 0 to 128;
20
21  -- DSP builder top level.
22  component torusBreakdown_NonlinearSubsystem is
23    port (
24        In_a : in std_logic_vector(31 downto 0);
25        In_alpha : in std_logic_vector(31 downto 0);
26        In_b : in std_logic_vector(31 downto 0);
```

```
27            In_beta : in std_logic_vector(31 downto 0);
28            In_dt : in std_logic_vector(31 downto 0);
29            In_x : in std_logic_vector(31 downto 0);
30            In_y : in std_logic_vector(31 downto 0);
31            In_z : in std_logic_vector(31 downto 0);
32            Out_x : out std_logic_vector(31 downto 0);
33            Out_y : out std_logic_vector(31 downto 0);
34            Out_z : out std_logic_vector(31 downto 0);
35            clk : in std_logic;
36            areset : in std_logic;
37            h_areset : in std_logic);
38      end component;
39      -- END DSP builder top level.
40
41      -- latency : 5 clock cycles (scale for DAC range)
42      component floatingPointMultiplyDedicated IS
43         PORT
44         (
45            aclr : IN STD_LOGIC ;
46            clock : IN STD_LOGIC ;
47            dataa : IN STD_LOGIC_VECTOR (31 DOWNTO 0);
48            datab : IN STD_LOGIC_VECTOR (31 DOWNTO 0);
49            result : OUT STD_LOGIC_VECTOR (31 DOWNTO 0)
50         );
51      END component;
52
53      -- latency : 6 clock cycles
54      component floatingPointToFixed IS
55         PORT
56         (
57            aclr : IN STD_LOGIC ;
58            clock : IN STD_LOGIC ;
59            dataa : IN STD_LOGIC_VECTOR (31 DOWNTO 0);
60            result : OUT STD_LOGIC_VECTOR (31 DOWNTO 0)
61         );
62      END component;
63
64      component floatingPointAddSubtract IS
65         PORT
66         (
67            aclr : IN STD_LOGIC ;
68            add_sub : IN STD_LOGIC ;
69            clk_en : IN STD_LOGIC ;
70            clock : IN STD_LOGIC ;
71            dataa : IN STD_LOGIC_VECTOR (31 DOWNTO 0);
72            datab : IN STD_LOGIC_VECTOR (31 DOWNTO 0);
73            result : OUT STD_LOGIC_VECTOR (31 DOWNTO 0)
74         );
75      END component;
76
77      component pulseFSM is port (
78            reset,clock,trigger : in std_logic;
79            pulseOut,pulseOutSingleClockCycle : out std_logic);
```

```
80   end component;
81
82   compcnent dFlipFlopWithAsyncReset is port (
83      clock,reset : in std_logic;
84      d,resetVal : in std_logic_vector(31 downto 0);
85      q : out std_logic_vector(31 downto 0));
86   end component;
87
88   begin
89      reset <= not resetn;
90      -- Euler's method
91      -- We first synchronously update state variables at
            781.250 KHz (64 counts of 50 MHz clock)
92      -- Since dt = 1/1024, time scale is actually (780.250e3
            /1024) = 762 Hz (approximately)
93
94      -- since synchronous update count is integer, simply use
            a process statement
95      process(incrementCountPulseN,resetn)
96      begin
97         if resetn = '0' then
98             countIncrement <= 64;
99         else
100            if falling_edge(incrementCountPulseN) then
101                if incrementCount = '1' then
102                    countIncrement <= countIncrement+1;
103                else
104                    countIncrement <= countIncrement-1;
105                end if;
106            end if;
107        end if;
108     end process;
109     -- state memory
110     process(clockIn, resetn)
111     begin
112            -- constants (place outside reset and clock to
                  avoid latches)
113            dt <= X"3A800000"; -- 1/1024
114            -- default values for parameters and synchronous
                  count
115            beta <= X"3F800000"; -- 1
116            outputScaledBy4 <= X"40800000";
117            a <= X"3D8F5C28"; -- 0.07, approximately 0.069999
118            b <= X"3DCCCCCC"; -- 0.1, approximately 0.099999
119         if resetn = '0' then
120            -- initial state
121            x <= X"3DCCCCCC"; -- 0.1
122            y <= X"00000000"; -- 0
123            z <= X"3DCCCCCC"; -- 0.1
124            count <= 0;
125         else
126             if rising_edge(clockIn) then
127                 if count = countIncrement then
```

```
128                         count <= 0;
129                     else
130                         count <= count + 1;
131                     end if;
132
133                     if count = countIncrement-1 then
134                         x <= xNew;
135                         y <= yNew;
136                         z <= zNew;
137                     end if;
138                 end if;
139             end if;
140     end process;
141
142     incrementAlphaPulse <= not incrementAlphaPulseN;
143     pulseFSMForAlpha : pulseFSM port map (
144         reset => reset,
145         clock => clockIn,
146         trigger => incrementAlphaPulse,
147         pulseOut => incrementDecrementAlphaPulse,
148         pulseOutSingleClockCycle => dFlipFlopClock);
149
150     alphaParameterBifurcation : floatingPointAddSubtract port
            map (
151         aclr => reset,
152         add_sub => incrementAlpha, -- '1' = add, '0' =
                subtract
153         clk_en => incrementDecrementAlphaPulse,
154         clock => clockIn,
155         dataa => alpha, -- start at 15. Memory implemented
                using D flip-flop
156         datab => X"3DCCCCCC", -- increment/decrement by 0.1 =
                X"3DCCCCCC"
157         result => alphaSignal);
158
159 -- we will need the flip-flop below to provide a proper
        initial state.
160     alphaFlipFlop : dFlipFlopWithAsyncReset port map (
161         clock => dFlipFlopClock,
162         reset => reset,
163         d => alphaSignal,
164         resetVal => X"41700000", -- start at 15
165         q => alpha);
166 --
167     staticNonlinearitiesAndDeltaT :
            torusBreakdown_NonlinearSubsystem port map (
168         In_a => a,
169         In_alpha => alpha,
170         In_b => b,
171          In_beta => beta,
172         In_dt => dt,
173         In_x => x,
174         In_y => y,
```

```vhdl
175         In_z => z,
176         Out_x => xNew,
177         Out_y => yNew,
178         Out_z => zNew,
179         clk => clockIn,
180         areset => reset,
181         h_areset => reset);
182    -- END Euler's method
183
184    -- scale outputs
185    scaleX : floatingPointMultiplyDedicated port map (
186        aclr => reset,
187        clock => clockIn,
188        dataa => x,
189        datab => outputScaledBy4,
190        result => xScaled);
191    scaleY : floatingPointMultiplyDedicated port map (
192        aclr => reset,
193        clock => clockIn,
194        dataa => y,
195        datab => outputScaledBy4,
196        result => yScaled);
197    scaleZ : floatingPointMultiplyDedicated port map (
198        aclr => reset,
199        clock => clockIn,
200        dataa => z,
201        datab => outputScaledBy4,
202        result => zScaled);
203
204    --state outputs : convert scaled floating point x,y
                 variables to 2.30 fixed point for DAC
205    xOutFinal : floatingPointToFixed port map (
206        aclr => reset,
207        clock => clockIn,
208        dataa => xScaled,
209        result => xFixed);
210    yOutFinal : floatingPointToFixed port map (
211        aclr => reset,
212        clock => clockIn,
213        dataa => yScaled,
214        result => yFixed);
215    zOutFinal : floatingPointToFixed port map (
216        aclr => reset,
217        clock => clockIn,
218        dataa => zScaled,
219        result => zFixed);
220
221        xOut <= xFixed(31 downto 16);
222        yOut <= yFixed(31 downto 16);
223        zOut <= zFixed(31 downto 16);
224
225 end behavioral;
```

Appendix F
Chapter 5 VHDL

F.1 Flip-Flops in VHDL

Listing F.1 VHDL specification of D flip-flop with asynchronous reset

```
 1  library ieee;
 2  use ieee.std_logic_1164.all;
 3
 4  entity dFlipFlopWithAsyncReset is port (
 5     clock,reset : in std_logic;
 6     d : in std_logic_vector(31 downto 0);
 7     q : out std_logic_vector(31 downto 0));
 8  end dFlipFlopWithAsyncReset;
 9
10  architecture structuralDFlipFlop of dFlipFlopWithAsyncReset
       is
11
12  begin
13     process (clock,reset)
14     begin
15        if reset = '1' then
16           q <= X"00000000";
17        else
18           if rising_edge(clock) then
19              q <= d;
20           end if;
21        end if;
22     end process;
23
24  end structuralDFlipFlop;
```

© Springer International Publishing Switzerland 2015
B. Muthuswamy and S. Banerjee, *A Route to Chaos Using FPGAs*, Emergence,
Complexity and Computation 16, DOI 10.1007/978-3-319-18105-9

F.2 VHDL Tapped Delay Line

Listing F.2 Realization of VHDL delay

```
 1  library ieee;
 2  use ieee.std_logic_1164.all;
 3  use ieee.numeric_std.all;
 4
 5  entity addressableShiftRegister is
 6          generic (numberOfFlipFlops : integer := 0;
 7                       delay : integer := 0);
 8          port (
 9          clk,areset : in std_logic;
10        In_x : in std_logic_vector(31 downto 0);
11            Out_xDelayed : out std_logic_vector(31 downto 0));
12  end addressableShiftRegister;
13
14  architecture behavioral of addressableShiftRegister is
15
16     component dFlipFlopWithAsyncRChap9eset is port (
17        clock,reset : in std_logic;
18        d : in std_logic_vector(31 downto 0);
19        q : out std_logic_vector(31 downto 0));
20     end component;
21
22     type memory is array(0 to numberOfFlipFlops) of
              std_logic_vector(31 downto 0);
23     signal internalDataArray : memory;
24
25  begin
26
27     internalDataArray(0) <= In_x;
28     generateFlipFlops:
29        for i IN 0 to numberOfFlipFlops-1 generate
30           nFlipFlops : dFlipFlopWithAsyncReset port map (
31              clock => clk,
32              reset => areset,
33              d => internalDataArray(i),
34              q => internalDataArray(i+1));
35        end generate;
36
37     Out_xDelayed <= internalDataArray(delay-1);
38  end behavioral;
```

F.3 VHDL Specification of Ikeda DDE

Listing F.3 Realization of Ikeda DDE

```
1   -- In order to implement a DDE, we need to implement two modules:
2   -- 1. An addressable shift register that implements the delay (refer
       to addressableShiftRegister subsystem)
3   -- 2. An appropriate clock for the addressable shift register, based
       on the global clock divider. The addressableShiftRegisterClock has
       2x the count of the clock
4   -- divider. Hence, the rising edge for the addressable shift register
       clock occurs whenever the 50 MHz clock divider count overflows.
5
6   library ieee;
7   use ieee.std_logic_1164.all;
8   use ieee.std_logic_signed.all;
9
10  entity ikedaDDESystem is port (
11        resetn, clockIn : in std_logic;
12        xFloatOut,xDelayedFloatOut : out std_logic_vector(31 downto 0);
13        xOut,xDelayedOut : out std_logic_vector(15 downto 0);
14        syncIn : in std_logic_vector(31 downto 0);
15        syncClock : out std_logic);
16  end ikedaDDESystem;
17
18  architecture behavioral of ikedaDDESystem is
19
20  signal reset : std_logic;
21  -- constants
22  signal dt : std_logic_vector(31 downto 0);
23  -- state variables
24  signal x,xDelayed,xNew,xFixed,xDelayedFixed : std_logic_vector(31
         downto 0);
25  -- prescalar
26  signal count: integer range 0 to 64;
27  signal addressableShiftRegisterCount : integer range 0 to 128;
28  signal addressableShiftRegisterClock : std_logic;
29  signal internalSyncClockCount : integer range 0 to 2048;
30  signal internalSyncClock : std_logic;
31
32  -- DSP builder top level. Add both the nonlinear subsystem and
         nonlinear synthesizable
33  -- subsystem VHDL files to project.
34  component ikedaDDE_NonlinearSubsystem is
35    port (
36        In_dt : in std_logic_vector(31 downto 0);
37         In_sync : in std_logic_vector(31 downto 0);
38        In_x : in std_logic_vector(31 downto 0);
39        In_xDelayed : in std_logic_vector(31 downto 0);
40        y : out std_logic_vector(31 downto 0);
41        clk : in std_logic;
42        areset : in std_logic;
43        h_areset : in std_logic);
44  end component;
45  -- END DSP builder top level.
46
47  component addressableShiftRegister is
```

```vhdl
48          generic (numberOfFlipFlops : integer := 0;
49                      delay : integer := 0);
50          port (
51          clk,areset : in std_logic;
52        In_x : in std_logic_vector(31 downto 0);
53        Out_xDelayed : out std_logic_vector(31 downto 0));
54  end component;
55
56
57  -- latency : 6 clock cycles
58  component floatingPointToFixed IS
59    PORT
60    (
61        aclr : IN STD_LOGIC ;
62        clock : IN STD_LOGIC ;
63        dataa : IN STD_LOGIC_VECTOR (31 DOWNTO 0);
64        result : OUT STD_LOGIC_VECTOR (31 DOWNTO 0)
65    );
66  END component;
67
68  begin
69      reset <= not resetn;
70      -- Euler's method
71      -- We first synchronously update state variables at 781.250 KHz (64
                counts of 50 MHz clock)
72      -- Since dt = 1/1024, time scale is actually (781.250e3/1024) = 762
                Hz (approximately)
73      -- state memory
74      process(clockIn, resetn)
75      begin
76              -- constants (place outside reset and clock to avoid latches
                    )
77              dt <= X"3A800000"; -- 1/1024
78          if resetn = '0' then
79              -- initial state
80              x <= X"3DCCCCCC";-- 0.1
81              count <= 0;
82              addressableShiftRegisterCount <= 0;
83              addressableShiftRegisterClock <= '0';
84          else
85              if rising_edge(clockIn) then
86
87                  if count = 64 then
88                      count <= 0;
89                  else
90                      count <= count + 1;
91                  end if;
92
93                  if count = 63 then
94                      x <= xNew;
95                  end if;
96
97                  if addressableShiftRegisterCount = 128 then
98                      addressableShiftRegisterCount <= 0;
99                  else
100                     addressableShiftRegisterCount <=
                            addressableShiftRegisterCount + 1;
```

```
101                        end if;
102
103                        if addressableShiftRegisterCount >= 63 then
104                            addressableShiftRegisterClock <= '1';
105                        else
106                            addressableShiftRegisterClock <= '0';
107                        end if;
108
109                        -- for synchronizer period
110                        if internalSyncClockCount = 2048 then
111                            internalSyncClockCount <= 0;
112                        else
113                            internalSyncClockCount < =
114                                internalSyncClockCount + 1;
115                        end if;
116
117                        if internalSyncClockCount >= 1023 then
118                            internalSyncClock <= '1';
119                        else
120                            internalSyncClock <= '0';
121                        end if;
122
123                    end if;
124            end if;
125        end process;
126
127    -- this design also includes synchronization
128    -- since y(t+dt)=y(t)+(-alpha*y+mu*sin(y(t-tau))+k(t)(x(t)-y(t)))*
           dt,
129    -- we also send in the sync signal into the DSP builder nonlinear
           subsystem.
130    staticNonlinearities : ikedaDDE_NonlinearSubsystem port map (
131        In_dt => dt,
132          In_sync => syncIn,
133        In_x => x,
134        In_xDelayed => xDelayed,
135        y => xNew,
136        clk => clockIn,
137        areset => reset,
138        h_areset => reset);
139
140
141    delay : addressableShiftRegister generic map (numberOfFlipFlops =>
           2048,delay => 1024)
142        port map (
143        In_x => x,
144        Out_xDelayed => xDelayed,
145        clk => addressableShiftRegisterClock,
146        areset => reset);
147    -- END Euler's method
148
149    --state outputs : convert scaled floating point variables to 5.27
           fixed point format DAC (no latency)
150    xOutFinal : floatingPointToFixed port map (
151        aclr => reset,
152        clock => clockIn,
153        dataa => x,
```

```
154          result => xFixed);
155     xDelayedOutFinal : floatingPointToFixed port map (
156          aclr => reset,
157          clock => clockIn,
158          dataa => xDelayed,
159          result => xDelayedFixed);
160
161          xOut <= xFixed(31 downto 16);
162          xDelayedOut <= xDelayedFixed(31 downto 16);
163
164          xFloatOut <= x;
165          xDelayedFloatOut <= xDelayed;
166          syncClock <= internalSyncClock;
167     end behavioral;
```

F.4 VHDL Specification of DDE with Sigmoidal Nonlinearity

Listing F.4 Realization of Sigmoidal DDE

```
1   library ieee;
2   use ieee.std_logic_1164.all;
3   use ieee.std_logic_signed.all;
4
5   entity sigmoidDDESystem is port (
6          resetn, clockIn : in std_logic;
7          xFloatOut,xDelayedFloatOut : out std_logic_vector(31 downto 0);
8          xOut,xDelayedOut : out std_logic_vector(15 downto 0);
9          syncIn : in std_logic_vector(31 downto 0);
10         syncClock : out std_logic);
11  end sigmoidDDESystem;
12
13  architecture behavioral of sigmoidDDESystem is
14
15  signal reset : std_logic;
16  -- constants
17  signal dt : std_logic_vector(31 downto 0);
18  -- state variables
19  signal x,xDelayed,xNew,xFixed,xDelayedFixed : std_logic_vector(31
            downto 0);
20  -- prescalar
21  signal count: integer range 0 to 64;
22  signal addressableShiftRegisterCount : integer range 0 to 128;
23  signal addressableShiftRegisterClock : std_logic;
24  signal internalSyncClockCount : integer range 0 to 2048;
25  signal internalSyncClock : std_logic;
26
27  -- DSP builder top level. Add both the nonlinear subsystem and
            nonlinear synthesizable
28  -- subsystem VHDL files to project.
29  component sigmoidalDDE_NonlinearSubsystem is
30      port (
31         In_dt : in std_logic_vector(31 downto 0);
32         In_sync : in std_logic_vector(31 downto 0);
```

```vhdl
33          In_x : in std_logic_vector(31 downto 0);
34          In_xDelayed : in std_logic_vector(31 downto 0);
35          y : out std_logic_vector(31 downto 0);
36          clk : in std_logic;
37          areset : in std_logic;
38          h_areset : in std_logic
39          );
40    end component;
41    -- END DSP builder top level.
42
43    component addressableShiftRegister is
44          generic (numberOfFlipFlops : integer := 0;
45                        delay : integer := 0);
46          port (
47          clk,areset : in std_logic;
48        In_x : in std_logic_vector(31 downto 0);
49          Out_xDelayed : out std_logic_vector(31 downto 0));
50    end component;
51
52
53    -- latency : 6 clock cycles
54    component floatingPointToFixed IS
55      PORT
56      (
57          aclr : IN STD_LOGIC ;
58          clock : IN STD_LOGIC ;
59          dataa : IN STD_LOGIC_VECTOR (31 DOWNTO 0);
60          result : OUT STD_LOGIC_VECTOR (31 DOWNTO 0)
61      );
62    END component;
63
64    begin
65      reset <= not resetn;
66      -- Euler's method
67      -- We first synchronously update state variables at 3.125 MHz (16
              counts of 50 MHz clock)
68      -- Since dt = 1/1000, time scale is actually (3.125e6/1000) = 3.125
              kHz (approximately)
69      -- state memory
70      process(clockIn, resetn)
71      begin
72              -- constants (place outside reset and clock to avoid latches
                  )
73              dt <= X"3A83126E"; -- 1/1000
74          if resetn = '0' then
75              -- initial state
76              x <= X"3DCCCCCC"; -- 0.1
77              count <= 0;
78              addressableShiftRegisterCount <= 0;
79              addressableShiftRegisterClock <= '0';
80          else
81              if rising_edge(clockIn) then
82
83                  if count = 16 then
84                      count <= 0;
85                  else
86                      count <= count + 1;
```

```
87                    end if;
88
89                    if count = 15 then
90                       x <= xNew;
91                    end if;
92
93                    if addressableShiftRegisterCount = 16 then
94                       addressableShiftRegisterCount <= 0;
95                    else
96                       addressableShiftRegisterCount <=
                            addressableShiftRegisterCount + 1;
97                    end if;
98
99                    if addressableShiftRegisterCount >= 7 then
100                      addressableShiftRegisterClock <= '1';
101                   else
102                      addressableShiftRegisterClock <= '0';
103                   end if;
104
105                   -- for synchronizer period
106                   if internalSyncClockCount = 2048 then
107                      internalSyncClockCount <= 0;
108                   else
109                      internalSyncClockCount < =
110                         internalSyncClockCount + 1;
111                   end if;
112
113                   if internalSyncClockCount >= 1023 then
114                      internalSyncClock <= '1';
115                   else
116                      internalSyncClock <= '0';
117                   end if;
118
119               end if;
120           end if;
121       end process;
122
123       -- this design also includes synchronization, so we also send in
              the sync signal into the DSP builder nonlinear subsystem.
124       staticNonlinearities : sigmoidalDDE_NonlinearSubsystem port map (
125           In_dt => dt,
126           In_sync => syncIn,
127           In_x => x,
128           In_xDelayed => xDelayed,
129           y => xNew,
130           clk => clockIn,
131           areset => reset,
132           h_areset => reset);
133
134       delay : addressableShiftRegister generic map (numberOfFlipFlops =>
              4096,delay => 3000)
135           port map (
136           In_x => x,
137           Out_xDelayed => xDelayed,
138           clk => addressableShiftRegisterClock,
139           areset => reset);
140       -- END Euler's method
```

```
141
142       --state outputs : convert scaled floating point variables to 5.27
               fixed point format DAC (no latency)
143       xOutFinal : floatingPointToFixed port map (
144          aclr => reset,
145          clock => clockIn,
146          dataa => x,
147          result => xFixed);
148       xDelayedOutFinal : floatingPointToFixed port map (
149          aclr => reset,
150          clock => clockIn,
151          dataa => xDelayed,
152          result => xDelayedFixed);
153
154          xOut <= xFixed(31 downto 16);
155          xDelayedOut <= xDelayedFixed(31 downto 16);
156
157          xFloatOut <= x;
158          xDelayedFloatOut <= xDelayed;
159          syncClock <= internalSyncClock;
160       end behavioral;
```

F.5   VHDL Specification of DDE with Signum Nonlinearity

Listing F.5   Realization of Signum DDE

```
1    library ieee;
2    use ieee.std_logic_1164.all;
3    use ieee.std_logic_signed.all;
4
5    entity signumDDESystem is port (
6          resetn, clockIn : in std_logic;
7          xFloatOut,xDelayedFloatOut : out std_logic_vector(31 downto 0);
8          xOut,xDelayedOut : out std_logic_vector(15 downto 0);
9          syncIn : in std_logic_vector(31 downto 0);
10         syncClock : out std_logic);
11   end signumDDESystem;
12
13   architecture behavioral of signumDDESystem is
14
15   signal reset : std_logic;
16   -- constants
17   signal dt : std_logic_vector(31 downto 0);
18   -- state variables
19   signal x,xDelayed,xNew,xFixed,xDelayedFixed,f1,f2,f : std_logic_vector
           (31 downto 0);
20   -- prescalar
21   signal count: integer range 0 to 64;
22   signal addressableShiftRegisterCount : integer range 0 to 128;
23   signal addressableShiftRegisterClock : std_logic;
24   signal internalSyncClockCount : integer range 0 to 2048;
25   signal internalSyncClock : std_logic;
26
```

```
27  -- We do not use DSP builder for the signum system since the signum
        function is very easy to implement:
28  -- from http://en.wikipedia.org/wiki/Single-precision_floating-
        point_format
29  -- sign bit is MSb, +0 = 0x00000000, -0 = 0x80000000
30  -- System: x'=sgn(x(t-2))-x(t-2), x(t <= 0) = 0.1
31  signal isZero,signbit : std_logic;
32  signal signbitOut : std_logic_vector(31 downto 0);
33
34  component addressableShiftRegister is
35          generic (numberOfFlipFlops : integer := 0;
36                       delay : integer := 0);
37          port (
38          clk,areset : in std_logic;
39        In_x : in std_logic_vector(31 downto 0);
40          Out_xDelayed : out std_logic_vector(31 downto 0));
41  end component;
42
43
44  -- latency : 6 clock cycles
45  component floatingPointToFixed IS
46     PORT
47     (
48        aclr : IN STD_LOGIC ;
49        clock : IN STD_LOGIC ;
50        dataa : IN STD_LOGIC_VECTOR (31 DOWNTO 0);
51        result : OUT STD_LOGIC_VECTOR (31 DOWNTO 0)
52     );
53  END component;
54
55  -- add_sub = '1' for addition, else subtraction
56  component floatingPointAddSubtract IS
57     PORT
58     (
59        aclr : IN STD_LOGIC ;
60        add_sub : IN STD_LOGIC ;
61        clk_en : IN STD_LOGIC ;
62        clock : IN STD_LOGIC ;
63        dataa : IN STD_LOGIC_VECTOR (31 DOWNTO 0);
64        datab : IN STD_LOGIC_VECTOR (31 DOWNTO 0);
65        result : OUT STD_LOGIC_VECTOR (31 DOWNTO 0)
66     );
67     END component;
68
69  component floatingPointMultiplyDedicated IS
70     PORT
71     (
72        aclr : IN STD_LOGIC ;
73        clock : IN STD_LOGIC ;
74        dataa : IN STD_LOGIC_VECTOR (31 DOWNTO 0);
75        datab : IN STD_LOGIC_VECTOR (31 DOWNTO 0);
76        result : OUT STD_LOGIC_VECTOR (31 DOWNTO 0)
77     );
78  END component;
79
80  begin
81     reset <= not resetn;
```

```
82      -- Euler's method
83      -- We first synchronously update state variables at 3.125 MHz (16
           counts of 50 MHz clock)
84      -- Since dt = 1/1000, time scale is actually (3.125e6/1000) = 3.125
           kHz (approximately)
85      -- state memory
86      process(clockIn, resetn)
87      begin
88              -- constants (place outside reset and clock to avoid
                    latches)
89            dt <= X"3A83126E"; -- 1/1000
90          if resetn = '0' then
91              -- initial state
92              x <= X"3DCCCCCC"; -- 0.1
93              count <= 0;
94              addressableShiftRegisterCount <= 0;
95              addressableShiftRegisterClock <= '0';
96          else
97              if rising_edge(clockIn) then
98
99                  if count = 16 then
100                     count <= 0;
101                 else
102                     count <= count + 1;
103                 end if;
104
105                 if count = 15 then
106                     x <= xNew;
107                 end if;
108
109                 if addressableShiftRegisterCount = 16 then
110                     addressableShiftRegisterCount <= 0;
111                 else
112                     addressableShiftRegisterCount <=
                            addressableShiftRegisterCount + 1;
113                 end if;
114
115                 if addressableShiftRegisterCount >= 7 then
116                     addressableShiftRegisterClock <= '1';
117                 else
118                     addressableShiftRegisterClock <= '0';
119                 end if;
120
121                 -- for synchronizer period
122                 if internalSyncClockCount = 2048 then
123                     internalSyncClockCount <= 0;
124                 else
125                     internalSyncClockCount <=
126                         internalSyncClockCount + 1;
127                 end if;
128
129                 if internalSyncClockCount >= 1023 then
130                     internalSyncClock <= '1';
131                 else
132                     internalSyncClock <= '0';
133                 end if;
134
```

```vhdl
135                  end if;
136            end if;
137      end process;
138
139      -- this design could include synchronization
140      delay : addressableShiftRegister generic map (numberOfFlipFlops =>
               4096,delay => 2000)
141          port map (
142          In_x => x,
143          Out_xDelayed => xDelayed,
144          clk => addressableShiftRegisterClock,
145          areset => reset);
146
147      -- compute f1=sgn(x(t-2))
148      with xDelayed select
149          isZero <= '1' when X"00000000",
150                        '1' when X"80000000",
151                        '0' when others;
152      signBit <= xDelayed(31);
153      with signBit select
154          signBitOut <= X"3F800000" when '0',  -- +1
155                          X"BF800000" when others; -- -1
156      with isZero select
157          f1 <= signBitOut when '0',
158                X"00000000" when others;
159
160      -- compute f2=sgn(x(t-2))-x(t-2)=f1-xDelayed
161      f2Out : floatingPointAddSubtract port map (
162          aclr => reset,
163          add_sub => '0',
164          clk_en => '1',
165          clock => clockIn,
166          dataa => f1,
167          datab => xDelayed,
168          result => f2);
169
170          -- compute f=f2*dt
171      fOut : floatingPointMultiplyDedicated port map (
172          aclr => reset,
173          clock => clockIn,
174          dataa => f2,
175          datab => dt,
176          result => f);
177      -- compute xNew = x+f
178      xNewOut : floatingPointAddSubtract port map (
179          aclr => reset,
180          add_sub => '1',
181          clk_en => '1',
182          clock => clockIn,
183          dataa => x,
184          datab => f,
185          result => xNew);
186      -- END Euler's method
187
188      --state outputs : convert scaled floating point variables to 5.27
                fixed point format DAC (no latency)
189      xOutFinal : floatingPointToFixed port map (
```

```
190          aclr => reset,
191          clock => clockIn,
192          dataa => x,
193          result => xFixed);
194     xDelayedOutFinal : floatingPointToFixed port map (
195          aclr => reset,
196          clock => clockIn,
197          dataa => xDelayed,
198          result => xDelayedFixed);
199
200          xOut <= xFixed(31 downto 16);
201          xDelayedOut <= xDelayedFixed(31 downto 16);
202
203          xFloatOut <= x;
204          xDelayedFloatOut <= xDelayed;
205          syncClock <= internalSyncClock;
206     end behavioral;
```

F.6 VHDL Specification for Chaotic DDE Synchronization

Listing F.6 VHDL module for implementing synchronization schemes

```
1    library ieee;
2    use ieee.std_logic_1164.all;
3    use ieee.numeric_std.all;
4
5    entity synchronizer is port (
6          resetn,clockIn,syncMode,xMinusYIn : in std_logic;
7          syncClock : in std_logic;
8          xIn,yIn,yDelayedIn : in std_logic_vector(31 downto 0);
9          syncOut : out std_logic_vector(31 downto 0));
10   end synchronizer;
11
12   architecture synchronizationSystem of synchronizer is
13       -- add_sub = '1' for addition, else subtraction
14       component floatingPointAddSubtract IS
15       PORT
16       (
17          aclr : IN STD_LOGIC ;
18          add_sub : IN STD_LOGIC ;
19          clk_en : IN STD_LOGIC ;
20          clock : IN STD_LOGIC ;
21          dataa : IN STD_LOGIC_VECTOR (31 DOWNTO 0);
22          datab : IN STD_LOGIC_VECTOR (31 DOWNTO 0);
23          result : OUT STD_LOGIC_VECTOR (31 DOWNTO 0)
24       );
25       END component;
26
27       component floatingPointMultiplyDedicated IS
28       PORT
```

```vhdl
29      (
30          aclr : IN STD_LOGIC ;
31          clock : IN STD_LOGIC ;
32          dataa : IN STD_LOGIC_VECTOR (31 DOWNTO 0);
33          datab : IN STD_LOGIC_VECTOR (31 DOWNTO 0);
34          result : OUT STD_LOGIC_VECTOR (31 DOWNTO 0)
35      );
36      END component;
37
38      component floatingPointCos IS
39      PORT
40      (
41          aclr : IN STD_LOGIC ;
42          clock : IN STD_LOGIC ;
43          data : IN STD_LOGIC_VECTOR (31 DOWNTO 0);
44          result : OUT STD_LOGIC_VECTOR (31 DOWNTO 0)
45      );
46      END component;
47
48      component floatingPointAbs IS
49      PORT
50      (
51          aclr : IN STD_LOGIC ;
52          data : IN STD_LOGIC_VECTOR (31 DOWNTO 0);
53          result : OUT STD_LOGIC_VECTOR (31 DOWNTO 0)
54      );
55      END component;
56
57      signal reset : std_logic;
58      signal xMinusY,yMinusX,kTimesXMinusY,kTimesYMinusX,
            kSquareWaveTimesXMinusY,kSquareWaveTimesYMinusX,
            kCosineTimesXMinusY,kCosineTimesYMinusX :
            std_logic_vector(31 downto 0);
59      signal cosineOut,absOut,k1,cosCoupling : std_logic_vector
            (31 downto 0);
60      signal k,alpha,twoTimesMu : std_logic_vector(31 downto 0)
            ;
61  begin
62      reset <= not resetn;
63      alpha <= X"40A00000"; -- alpha = 5
64      twoTimesMu <= X"42200000"; -- two*mu = 2*20 = 40
65      -- generate x-y and y-x using megaWizard. We chose to use
            the megaWizard because DSP builder is too much effort
            for something as simple as floating point subtraction
66      xMinusYInstance : floatingPointAddSubtract port map (
67          aclr => reset,
68          add_sub => '0', -- dataa - datab
69          clk_en => '1',
70          clock => clockIn,
71          dataa => xIn,
72          datab => yIn,
```

```vhdl
73          result => XMinusY);
74
75      yMinusXInstance : floatingPointAddSubtract port map (
76          aclr => reset,
77          add_sub => '0', -- dataa - datab
78          clk_en => '1',
79          clock => clockIn,
80          dataa => yIn,
81          datab => xIn,
82          result => YMinusX);
83
84      -- generate square wave based on delayed clock
85      with syncClock select
86          k <= X"00000000" when '0',
87                  X"42480000" when others; -- square wave
88                       amplitude is 50
89      -- generate k(t) = -alpha+2*mu*|cos(y(t-tau))|
90      -- compute cosine
91      cosineYDelay : floatingPointCos port map (
92          aclr => reset,
93          clock => clockIn,
94          data => yDelayedIn,
95          result => cosineOut);
96      -- compute abs
97      absOfCos : floatingPointAbs port map (
98          aclr => reset,
99          data => cosineOut,
100         result => absOut);
101     -- compute product: 2*mu*|cos(y(t-tau))|
102     cosCouplingProductTerm : floatingPointMultiplyDedicated
103         port map (
104         aclr => reset,
105         clock => clockIn,
106         dataa => twoTimesMu,
107         datab => absOut,
108         result => k1);
109     -- subtract from alpha
110     cosineCouplingOut : floatingPointAddSubtract port map (
111         aclr => reset,
112         add_sub => '0', -- dataa - datab
113         clk_en => '1',
114         clock => clockIn,
115         dataa => k1,
116         datab => alpha,
117         result => cosCoupling);
118
119     -- generate k*(x-y), k*(y-x) (k is square wave)
120     kSquareWaveTimesXMinusYInstance :
            floatingPointMultiplyDedicated port map (
            aclr => reset,
```

```vhdl
121          clock => clockIn,
122          dataa => k,
123          datab => XMinusY,
124          result => kSquareWaveTimesXMinusY);
125
126      kSquareWaveTimesYMinusXInstance :
             floatingPointMultiplyDedicated port map (
127          aclr => reset,
128          clock => clockIn,
129          dataa => k,
130          datab => YMinusX,
131          result => kSquareWaveTimesYMinusX);
132
133      -- generate k*(x-y), k*(y-x) (k is cosine coupling)
134      kCosineTimesXMinusYInstance :
             floatingPointMultiplyDedicated port map (
135          aclr => reset,
136          clock => clockIn,
137          dataa => k,
138          datab => XMinusY,
139          result => kCosineTimesXMinusY);
140
141      kCosineTimesYMinusXInstance :
             floatingPointMultiplyDedicated port map (
142          aclr => reset,
143          clock => clockIn,
144          dataa => k,
145          datab => YMinusX,
146          result => kCosineTimesYMinusX);
147
148      -- syncMode: '0' - square wave coupling, '1' - cosine
                coupling
149      with syncMode select
150          kTimesXMinusY <= kSquareWaveTimesXMinusY when '0',
151                                   kCosineTimesXMinusY when
                                            others;
152
153      with syncMode select
154          kTimesYMinusX <= kSquareWaveTimesYMinusX when '0',
155                                   kCosineTimesYMinusX when
                                            others;
156      with xMinusYIn select
157          syncOut <= kTimesXMinusY when '1',
158                           kTimesYMinusX when others;
159
160  end synchronizationSystem;
```

F.7 DE2 Chaotic DDE Specification Top Level

Listing F.7 VHDL top level for implementing synchronization scheme for DDEs

```
1    -- DE2-115 Audio codec interface for analog chaotic signals output
         from chaotic DDEs
2    -- Reference design for EPJ-ST paper - "Synchronization in Coupled
         Ikeda Delay Differential Equations : Experimental Observations
         using Field
3    -- Programmable Gate Arrays", Valli, D. et. al.
4    -- i2c Audio codec interface courtesy of Stapleton, Colin, EE2902
         Spring 2011, Milwaukee School of Engineering, 4/15/11
5    -- adc_dac interface courtesy of Chu, Embedded SOPC Design with
         VHDL, pp. 545-546, 4/8/13
6    -- Based on chaos engine created by Dr. Muthuswamy (AY 2013-2014),
         primarily for the book:
7    -- "A Route to Chaos Using FPGAs - Volumes I and II"
8    -- DSP builder based design
9    -- IMPORTANT : NEED TO SET nCEO as regular I/O via Device and Pin
         Options
10   -- KEY(0): global reset
11   -- KEY(1): bifurcation parameter control (see SW(2) below), but
         this is not used in this design.
12   -- SW(0) : loopback (switch down or SW(0)='0') or chaotic dynamics
         (switch up or SW(0)='1' (default))
13   -- SW(1) : select between drive (switch down or SW(0)='0') and sync
         error signal (switch up or SW(0)='1', e(t) for square wave is
         on left-channel. e(t) for cosine on right channel)
14   -- SW(2) : bifurcation parameter decrement (switch down or SW(2) =
         '0') or increment (switch up or SW(2) = '1'
15   (default)). This is not used in this design.
16   -- left channel data is controlled by SW(4 downto 3)
17   -- right channel data is controlled by SW(6 downto 5)
18
19
20   LIBRARY ieee;
21   USE ieee.std_logic_1164.all;
22   USE ieee.std_logic_signed.all; -- need to subtract
         std_logic_vectors for synchronization result
23
24   entity DE2ChaoticDDEs is
25     port(
26        KEY: in std_logic_vector(3 downto 0);
27        CLOCK_50: in std_logic;
28        --I2C ports
29        I2C_SCLK: out std_logic;
30        I2C_SDAT: inout std_logic;
31        --audio codec ports
32        AUD_ADCDAT: in std_logic;
33        AUD_ADCLRCK: out std_logic;
34        AUD_DACLRCK: out std_logic;
35        AUD_DACDAT: out std_logic;
36        AUD_XCK: out std_logic;
37        AUD_BCLK: out std_logic;
38        --select loopback test or neuron model output
```

```vhdl
39          SW: in std_logic_vector(17 downto 0);
40          --output for logic analyzer
41          GPIO: inout std_logic_vector (35 downto 0);
42          HEX3,HEX2,HEX1,HEX0: out std_logic_vector(6 downto 0);
43          LEDG: out std_logic_vector (8 downto 0)
44      );
45  end DE2ChaoticDDEs;
46
47  architecture toplevel of DE2ChaoticDDEs is
48
49      --PLL from MegaWizard in Quartus.
50      --both input and output are 50MHz
51      component clockBuffer IS
52        PORT
53        (
54            areset : IN STD_LOGIC := '0';
55            inclk0 : IN STD_LOGIC := '0';
56            c0 : OUT STD_LOGIC
57        );
58      END component;
59
60      --I2C controller to drive the Wolfson codec
61      component audioCodecController is
62      port(
63          clock50MHz,reset: in std_logic;
64          I2C_SCLK_Internal: out std_logic;
65          --must be inout to allow FPGA to read the ack bit
66          I2C_SDAT_Internal: out std_logic;
67          SDAT_Control: out std_logic;
68          --for testing
69          clock50KHz_Out: out std_logic
70      );
71      end component;
72
73      --generates digital audio interface clock signals
74      component adc_dac is
75      port (
76        clk, reset: in std_logic; -- reset signal starts '0' then goes
                  to '1' after 40 ms => active low. Hence we will complement
                  the delayed reset signal
77        dac_data_in: in std_logic_vector(31 downto 0);
78        adc_data_out: out std_logic_vector(31 downto 0);
79        m_clk, b_clk, dac_lr_clk, adc_lr_clk: out std_logic;
80        dacdat: out std_logic;
81        adcdat: in std_logic;
82        load_done_tick: out std_logic
83      );
84      end component;
85
86      component topLevelMux is port (
87          loopbackN,chaosSystemSelect : in std_logic;
88          selectLeftChannelBits,selectRightChannelBits : in
                  std_logic_vector(1 downto 0);
89          leftChannelLoopBackIn,rightChannelLoopBackIn: in
                  std_logic_vector(15 downto 0);
```

```
90          chaosX1,chaosY1,chaosZ1,chaosX2,chaosY2,chaosZ2 : in
               std_logic_vector(15 downto 0);
91          leftChannelDACRegister,rightChannelDACRegister : out
               std_logic_vector(15 downto 0));
92      end component;
93
94      component pulseFSM is port (
95          reset,clock,trigger : in std_logic;
96          pulseOut : out std_logic);
97      end component;
98
99      -- Chaotic system. DSP builder static nonlinearity interface is
               inside the module.
100     -- The ikeda DDE subsystem can also implement synchronization
101     component ikedaDDESystem is port (
102         resetn, clockIn : in std_logic;
103         xFloatOut,xDelayedFloatOut : out std_logic_vector(31 downto
               0);
104         xOut,xDelayedOut : out std_logic_vector(15 downto 0);
105         syncIn : in std_logic_vector(31 downto 0);
106         syncClock : out std_logic);
107     end component;
108
109     component sigmoidDDESystem is port (
110         resetn, clockIn : in std_logic;
111         xFloatOut,xDelayedFloatOut : out std_logic_vector(31 downto
               0);
112         xOut,xDelayedOut : out std_logic_vector(15 downto 0);
113         syncIn : in std_logic_vector(31 downto 0);
114         syncClock : out std_logic);
115     end component;
116
117     component signumDDESystem is port (
118         resetn, clockIn : in std_logic;
119         xFloatOut,xDelayedFloatOut : out std_logic_vector(31 downto
               0);
120         xOut,xDelayedOut : out std_logic_vector(15 downto 0);
121         syncIn : in std_logic_vector(31 downto 0);
122         syncClock : out std_logic);
123     end component;
124
125     component synchronizer is port (
126         resetn,clockIn,syncMode,xMinusYIn,syncClock : in std_logic;
127         xIn,yIn,yDelayedIn : in std_logic_vector(31 downto 0);
128         syncOut : out std_logic_vector(31 downto 0));
129     end component;
130     -- end dsp builder top level
131
132     --clock signal from the PLL clockBuffer
133     signal clock50MHz : std_logic;
134
135     --asynchronous reset for the whole project
136     signal reset: std_logic;
137
138     --I2C data and clock lines
```

```vhdl
139     signal i2cData, i2cClock: std_logic;
140
141     --tri-state buffer control
142     signal i2cDataControl: std_logic;
143     signal i2cDataTriState: std_logic;
144
145     --assert signal from delay counter
146     signal codecReset,codecResetn: std_logic;
147
148     --audio codec signals
149     signal clock18MHz : std_logic;
150     signal adcDat_sig: std_logic;
151     signal adcLRCK_sig: std_logic;
152     signal dacLRCK_sig: std_logic;
153     signal dacDat_sig: std_logic;
154     signal bck_sig: std_logic;
155     signal dac_data_in,adc_data_out : std_logic_vector(31 downto 0);
156
157     --nonlinear dynamics model signals
158     signal leftChannelDataRegister,leftChannelADCRegister,
            leftChannelDACRegister : std_logic_vector(15 downto 0);
159     signal rightChannelDataRegister,rightChannelADCRegister,
            rightChannelDACRegister : std_logic_vector(15 downto 0);
160     signal leftChannelChaos,rightChannelChaos : std_logic_vector(15
            downto 0);
161
162     signal chaosXOut,chaosYOut,chaosZOut : std_logic_vector(15
            downto 0);
163     signal chaosX1,chaosY1,chaosZ1 : std_logic_vector(15 downto 0);
164     signal chaosX2,chaosX2Temp,chaosY2,chaosZ2 : std_logic_vector(15
            downto 0);
165
166     -- 32-bit single-precision floating point data for
            synchronization
167     signal chaosXFloatOut,chaosYFloatOut,chaosXDelayedFloatOut,
            chaosYDelayedFloatOut,syncSignalXMinusY,syncSignalYMinusX :
            std_logic_vector(31 downto 0);
168     signal syncClock,syncClockTemp : std_logic;
169
170     signal trigger,pulseOut : std_logic;
171
172     -- testing
173     signal clock50KHz : std_logic;
174
175     -- select signals
176     signal selectBitsX,selectBitsY : std_logic_vector
177         (1 downto 0);
178 begin
179
180     --keys are active low
181     reset <= not KEY(0);
182
183     --PLL
184     clockBufferInstance: clockBuffer port map(reset,CLOCK_50,
            clock50MHz);
```

```
185
186      --I2C
187      I2CControllerInstance: audioCodecController port map(clock50MHz,
             reset, i2cClock, i2cData,i2cDataControl, clock50KHz);
188
189      --Codec Controller
190      dac_data_in <= leftChannelDACRegister&rightChannelDACRegister;
191      leftChannelADCRegister <= adc_data_out(31 downto 16);
192      rightChannelADCRegister <= adc_data_out(15 downto 0);
193      adcDacInterface : adc_dac port map (
194        clk => clock50MHz,
195          reset => reset,
196        dac_data_in => dac_data_in,
197        adc_data_out => adc_data_out,
198        m_clk=>clock18MHz,
199          b_clk => bck_sig,
200          dac_lr_clk => dacLRCK_sig,
201          adc_lr_clk => adcLRCK_sig,
202        dacdat => dacDat_sig,
203        adcdat => adcDat_sig,
204        load_done_tick => open);
205
206      mulitplexers : topLevelMux port map (
207      loopbackN => SW(0),
208      chaosSystemSelect => SW(1),
209      selectLeftChannelBits => SW(4 downto 3), -- SW(2) is used to
                 increment or decrement bifurcation parameter
210      selectRightChannelBits => SW(6 downto 5),
211      leftChannelLoopBackIn => leftChannelADCRegister,
212      rightChannelLoopBackIn => rightChannelADCRegister,
213      chaosX1 => chaosX1,
214      chaosY1 => chaosY1,
215      chaosZ1 => chaosZ1,
216      chaosX2 => chaosX2,
217      chaosY2 => chaosY2,
218      chaosZ2 => chaosZ2,
219      leftChannelDACRegister => leftChannelDACRegister,
220      rightChannelDACRegister => rightChannelDACRegister);
221
222
223      -- ----------------------- BEGIN DSP BUILDER BASED CHAOTIC
             DYNAMICS
224      -- IKEDA DDE SYNCHRONIZATION CURRENTLY DISABLED
225      -- ikedaDDESystemInstance_drive : ikedaDDESystem port map (
226      -- resetn => KEY(0),
227      -- clockIn => clock50MHz,
228      -- xFloatOut => chaosXFloatOut,
229      -- xDelayedFloatOut => chaosXDelayedFloatOut,
230      -- xOut => chaosX1,
231      -- xDelayedOut => chaosY1,
232      -- syncIn => X"000C0000", -- k(t)*(y-x)
233      -- syncClock => syncClock);
234      -- chaosZ1 <= X"0000";
235
236      -- ikedaDDESystemInstance_Response : ikedaDDESystem port map (
```

```
237   -- resetn => KEY(0),
238   -- clockIn => clock50MHz,
239   -- xFloatOut => chaosYFloatOut,
240   -- xDelayedFloatOut => chaosYDelayedFloatOut,
241   -- xOut => chaosX2,
242   -- xDelayedOut => chaosY2,
243   -- syncIn => syncSignalXMinusY, -- k(t)*(x-y)
244   -- syncClock => syncClockTemp);
245   -- --chaosX2 <= chaosX1 - chaosX2Temp; -- synchronization error
246   -- chaosZ2 <= X"0000";
247
248   --
249   -- syncInstanceSquareXMinusY : synchronizer port map (
250   -- resetn => KEY(0),
251   -- clockIn => clock50MHz,
252   -- syncMode => '0',
253   -- xMinusYIn => '1',
254   -- syncClock => syncClock,
255   -- xIn => chaosXFloatOut,
256   -- yIn => chaosYFloatOut,
257   -- yDelayedIn => chaosYDelayedFloatOut,
258   -- syncOut => syncSignalXMinusY);
259   --
260   -- syncInstanceSquareYMinusX : synchronizer port map (
261   -- resetn => KEY(0),
262   -- clockIn => clock50MHz,
263   -- syncMode => '0',
264   -- xMinusYIn => '0',
265   -- syncClock => syncClock,
266   -- xIn => chaosXFloatOut,
267   -- yIn => chaosYFloatOut,
268   -- yDelayedIn => chaosYDelayedFloatOut,
269   -- syncOut => syncSignalYMinusX);
270   --
271
272         sigmoidalDDE : sigmoidDDESystem port map (
273            resetn => KEY(0),
274            clockIn => clock50MHz,
275            xFloatOut => chaosXFloatOut, -- unused
276            xDelayedFloatOut => chaosXDelayedFloatOut,
277            xOut => chaosX1,
278            xDelayedOut => chaosY1,
279            syncIn => X"00000000",
280            syncClock => syncClock);
281         chaosZ1 <= X"0000";
282
283         signumDDE : signumDDESystem port map (
284            resetn => KEY(0),
285            clockIn => clock50MHz,
286            xFloatOut => chaosYFloatOut, -- unused
287            xDelayedFloatOut => chaosYDelayedFloatOut,
288            xOut => chaosX2,
289            xDelayedOut => chaosY2,
290            syncIn => X"00000000",
291            syncClock => syncClockTemp);
```

```
292              chaosZ2 <= X"0000";
293
294        ------------------------- END DSP BUILDER CHAOTIC DYNAMICS
295
296        --tri-state data output
297        i2cDataTriState <= i2cData when i2cDataControl = '1' else 'Z';
298
299        --I2C output ports
300        I2C_SCLK <= i2cClock;
301        I2C_SDAT <= i2cDataTriState;
302
303        --audio codec input port
304        adcDat_sig <= AUD_ADCDAT;
305
306        --audio codec ouput ports
307        AUD_ADCLRCK <= adcLRCK_sig;
308        AUD_DACLRCK <= dacLRCK_sig;
309        AUD_DACDAT <= dacDat_sig;
310        AUD_XCK <= clock18MHz;
311        AUD_BCLK <= bck_sig;
312
313        --for testing
314        GPIO(5) <= adcLRCK_sig;
315        GPIO(4) <= dacLRCK_sig;
316        GPIO(3) <= bck_sig;
317        GPIO(2) <= adcDat_sig;
318        GPIO(1) <= dacDat_sig;
319        GPIO(0) <= clock18MHz;
320        LEDG(0) <= reset;
321        LEDG(1) <= clock50KHz;
322
323
324        HEX1 <= "1111111";
325        HEX0 <= "1111111";
326
327        HEX2 <= "1111111";
328        HEX3 <= "1111111";
329
330  end toplevel;
```

The synchronizer functionality for the Ikeda DDE is implemented in Lines 222–267 in listing F.7. We have commented out the functionality so that the reader can experiment with synchronization.

1. Lines 222–244 define the Ikeda drive and response system.
2. Lines 247–267 implement two synchronizer modules : one for square wave coupling and the other for cosine coupling. Depending on the syncMode input, either square wave or cosine coupling is selected (refer to listing F.6).

Within each DDE DSP builder specification, we have included a "syncIn" input that can be tied to ground for non-synchronous designs.

Glossary

ADC Hardware component for interfacing external analog signals to digital platforms.

DAC Hardware component for converting digital into analog signals. They are usually at the output-end of digital systems and can be used, for example, to drive actuators.

DDE Differential equations that incorporate time delays. They usually need a continuum of initial conditions.

DSP These are digital integrated circuits that are designed for computation-intensive signal processing applications.

FPGA A massively parallel semiconductor device that can be "programmed" after manufacturing. The word programmed is in quotes since one should not think of designing for an FPGA as programming. Rather, one should think about what hardware one wants to specify for time critical execution on an FPGA. FPGAs are so named because they can be "programmed" in the field.

FSM Concept from automata theory that is used to describe sequential logic.

HDL Computer languages for specifying physical hardware. VHDL and Verilog are very popular; there are HDLs such as SystemC that is a set of C++ classes for event-driven simulation.

LAB Logical Array Blocks consist of groups of Logic Elements on Cyclone IV FPGAs.

LE Logic Element is the smallest unit of logic on a Cyclone IV FPGA.

LUT Look Up Tables map inputs to outputs and consist of 4 inputs in a Cyclone IV FPGa. Hence, we can specify $2^{16} = 65536$ logic functions in one LUT.

ODE Ordinary Differential Equation is an equation that consists of a function of one independent variable and its derivatives.

PLL These are nonlinear control systems that generate an output whose phase is locked with the input. We primarily use these in this book to buffer clock signals and as clock frequency multipliers.

ROM The data on these integrated circuits cannot be modified by the user, unlike random access memories.

© Springer International Publishing Switzerland 2015
B. Muthuswamy and S. Banerjee, *A Route to Chaos Using FPGAs*, Emergence, Complexity and Computation 16, DOI 10.1007/978-3-319-18105-9

RTL A design abstraction that models a synchronous digital circuit in terms of data flow between hardware registers and logic operations.

SDC A language for specifying timing constraints.

SDRAM A type of dynamic random access memory that is synchronized with a system bus.

SMA A type of coaxial connector used for frequencies typically from 0 to 18 GHz.

VHDL Hardware description language that can also be used for mixed-signal system specification and a general purpose parallel programming language.

Solutions

For step-by-step solutions to all problems, please visit the book's website:
http://www.harpgroup.org/muthuswamy/ARouteToChaosUsingFPGAs/ARouteTo
ChaosUsingFPGAs.html

© Springer International Publishing Switzerland 2015
B. Muthuswamy and S. Banerjee, *A Route to Chaos Using FPGAs*, Emergence,
Complexity and Computation 16, DOI 10.1007/978-3-319-18105-9

Printed in the United States
By Bookmasters